Appalachian
Winter

Appalachian Winter

MARCIA BONTA

University of Pittsburgh Press

Published by the University of Pittsburgh Press, Pittsburgh, Pa., 15260
Copyright © 2005, University of Pittsburgh Press
Manufactured in the United States of America
Printed on acid-free paper
10   9   8   7   6   5   4   3   2   1

Portions of this book first appeared, in slightly different form, in the
*Altoona Mirror, American Forests, Birdwatcher's Digest, Hawk Mountain
News, Pennsylvania Game News,* and *The Gnatcatcher,* newsletter of the
Juniata Valley Audubon Society.

Library of Congress Cataloging-in-Publication Data
Bonta, Marcia, 1940–
Appalachian winter / Marcia Bonta.
p. cm.
Includes bibliographical references and index.
ISBN 0-8229-5862-7 (pbk. : alk. paper)
1. Natural history—Pennsylvania.   2. Winter—Pennsylvania.
3. Natural history—Appalachian Region.   4. Winter—
Appalachian Region.   I. Title.
QH105.P4B66 2004
508.748—dc22
                                                    2004019672

Contents

Acknowledgments

MANY PEOPLE have encouraged me over the years. I never would have submitted *Appalachian Spring* to the University of Pittsburgh Press if Bill Randour, then an employee of the Western Pennsylvania Conservancy, hadn't suggested that I try them. I am especially grateful to Fred Hetzel, the former director of the University of Pittsburgh Press, who took a chance on my work and published both *Appalachian Spring* and *Appalachian Autumn*. The present director, Cynthia Miller, has also been supportive of my seasons books, and I thank her and her able staff.

I have worked with a number of magazine editors during my almost three decades of writing, but none can compare to Bob Mitchell of the *Pennsylvania Game News*. To have the freedom to write about the natural world in my own style has been a gift that I deeply appreciate.

Several teachers of nature writing have encouraged me. Dr. Ian Marshall of Penn State Altoona and his colleague, wildlife biologist Dr. Carolyn Mahan, have been using my books and bringing their environmental studies students to our property for guided walks and readings for a decade. Dr. Robert Burkholder of Penn State University Park has also been supportive of my work, and Dr. Tom Lyon, with whom I have corresponded, has used my writing in his anthology *On Nature's Terms* and has inspired me, by his kind words about my work in *This Incomparable Land*, to soldier on when I'm feeling discouraged. So too has writer Chris Bolgiano, a dear friend who loves the Appalachian forest as much as I do.

Field biologists are busy people, but Dr. Joseph F. Merritt, small mammal biologist and former Resident Director of

Acknowledgments

Powdermill Nature Reserve, the field station of the Carnegie Museum of Natural History, who is now Distinguished Professor at the U.S. Air Force Academy, has always given me prompt answers to my queries. Much of the information on red-backed voles, short-tailed shrews, and flying squirrels resulted from his research.

Members of my family have also been helpful, both as subjects and as editors. My son Dave carefully edited this book before I submitted it and his poetic voice helped to make it better. I also appreciate the use of his poem "In the Ice Forest" that first appeared in his weblog, *Via Negativa*, on February 7, 2004 (http://neithernor.blogspot.com). My other two sons—Steve and Mark—have always been encouraging even though they live far from their boyhood home now. My granddaughter, Eva, as always, is a joy and an inspiration.

Last, but not least, I thank my husband Bruce. Without him, I never would have dared to be a writer, and I certainly never would have lived on our Appalachian mountain. We have been happily married since 1962 and it seems like yesterday and yet forever. It's been a wonderful life.

In Memoriam

HAROLD C. MYERS
To Dad, who has "crossed the bar."

MY FATHER thought of himself as "Iron Man Myers": he could drive a car for days at a time. Behind the wheel, he was invincible. One time he drove almost nonstop from Spartanburg, South Carolina, to Woodbury, New Jersey. Right through Baltimore on old U.S. 1 at four in the morning he drove with my mother talking nervously to keep him awake.

In 1955 Dad took his family and our big 1950 green Oldsmobile with him to Europe for a nine-month business trip. He drove his beloved car on roads built for the tiny European cars of the time—along the narrow, winding Amalfi Drive in Italy, up to the top of the St. Bernard Pass in Switzerland, and through a cow pasture in Germany searching for a rural baroque church that some guidebook indicated was worth a detour.

I was fifteen, the oldest of four siblings, and in charge of following the maps, which showed only some of the roads to the castles, cathedrals, gardens, and villages Dad wanted to visit. Dad was particularly challenged by driving his big car on the left on the British roads. Mom had to tell him when he could pass other cars, which was nerve-wracking for him since he liked to fulfill his image of being in charge.

The whole European experience was an amazing adventure for a man born during a blizzard in a poor, hard coal village above Mahanoy City, Pennsylvania, called Vulcan. His paternal grandfather, who lived with them, had been an outside mine foreman. His father was a blacksmith there. But when

Dad was four years old, his father moved his family to Potts-town, where he worked in a factory as a metal treating spe-cialist.

Dad returned to the mountains every summer by train to spend his vacations with his mother's youngest sister, Mary Dresch. She had had two years of normal school and taught high school English, the only sibling in either family with ed-ucation beyond high school. My grandmother had gone through the seventh grade, my grandfather the fifth.

Though my grandfather saw little use in a college educa-tion, my grandmother had different ideas. Sweet though she seemed to me, she had an iron will and she was determined that her three boys would go to college. When Dad gradu-ated from high school he worked at the factory with his fa-ther, but only until he earned enough money to go to college.

It was the height of the Great Depression. He went to Penn State and instead of becoming a landscape architect (he loved gardening) or a surgeon (another ambition he quelled), this liberal arts man who recited poetry, read history, and de-voured fiction became a chemical engineer because that was where the jobs were. That was when he made his bargain with God. Make me a success and I will contribute generously to the church and church-related charities. As he saw it, God kept His end of the bargain and Dad kept his.

Our home in New Jersey became a showplace. He built stone walls and transplanted mountain laurel, rhododen-drons, and azaleas by the dozens to the oak woods that sur-rounded our home. Beyond our neighborhood stretched a series of small lakes and a forest—not a bad place to raise a family.

But it wasn't Pennsylvania and it wasn't the mountains, so we spent most of our vacations in Pottstown hiking the nearby hills where Dad had hiked as a boy. We often returned to Mahanoy City, especially on Memorial Day, to visit Aunt Mary and the family graves in the German Protestant cemetery near Vulcan Village on top of the mountain. It was during those visits that the ridge-and-valley section of Pennsylvania became my "mountains of home" just as they had for Dad.

Shortly after my husband Bruce and I and our three small sons moved to our mountain thirty-two years ago, Dad retired, sold his home, and started all over again in a country home near State College. Behind his house was a mountain, an extension of the same ridge we lived on twenty miles away. Tucked in the woods as his first home had been, it too was an ideal place to build stone walls and plant flowers and shrubs and that is what he did for more than two decades.

Mom died when he was eighty and he immediately asked Bruce and me to include him on our "outbound journeys in Pennsylvania," excursions we took for a column I wrote for seventeen years for the now-defunct *Pennsylvania Wildlife*. Gamely he hiked even in the snow, and we took him to as many places as we could so he could relive, in some sense, his childhood when he had roamed the hills around Pottstown. He also visited us frequently, driving up our road in his low-slung Oldsmobile and, inspired by the flowing stream that parallels the road, recited Alfred Lord Tennyson's "The Brook" each time.

We failed to see the signs of "Iron Man Myers" slowing down even when he had a couple near-misses in his car. Before we were forced to take away his keys, he fell and broke his hip. He refused to live with any of his children because he remembered how his mother had resented her father-in-law living with them for the first thirty years of their marriage, but he finally resigned himself to an assisted-living facility. We still brought him out to our place and took him for rides along country roads and through the forest, especially when the mountain laurel and rhododendron were in bloom.

One cloudy, threatening day I took him for a drive to see the autumn color in Rothrock State Forest near State College, where he had spent many happy hours hiking when he was a student at Penn State. Suddenly the sun broke through the clouds and the leaves glowed.

"I guess God knew that two nature lovers were out," he said happily. We saw no one else on those gravel roads and I drove fifteen miles an hour so he could identify the trees and shrubs. Dad and I also shared a love of operettas and I sang Sigmund Romberg's "Golden Days" as we drove along. He was so happy, reminiscing and "oohing" and "aahing" over the spectacular colors. Autumn had always been his favorite season, and he frequently recited all eight stanzas of Helen Hunt Jackson's poem "October's Bright Blue Weather."

He had an indomitable will and survived two operations for two broken hips. Each time he suffered through the pain and physical therapy to get back on his feet. But he also had severe spinal stenosis and after using a walker for several years he finally gave in to a wheelchair. We continued to bring him out to our place and once he said to me, "I'm always afraid this will be my last visit."

In the spring of 2003 he visited us for the last time. He was clearly in pain but he sat out on the veranda and looked at the field and forest that he loved so well. He ate his favorite meal of chili and enjoyed visiting with members of our extended family. After that, our only excursions were to a series of doctors about a series of problems. Yet whenever we visited he always minimized his pain and asked a cheery, "How's it with you?" And then, in late July, he died suddenly of pneumonia.

Now, whenever I walk down our road, I too recite "The Brook." "For men may come, and men may go, but I go on forever." Farewell, Dad. In all of our hearts, *you* will go on forever.

Introduction

I am convinced it is a most ridiculous thing
to go round the world when by staying quietly,
the world will go round with you.

Charles Darwin

WHEN I STARTED THIS SERIES on Appalachian seasons, I was a comparatively young woman. If I was lucky, my life would be less than half over. Now, I am at the cusp of elderhood, and every season is even more precious to me than it was when life seemed to stretch on forever. I am still able to walk our mountain trails and observe nature's processes, but I am keenly aware that keeping fit becomes more difficult every year. Every twinge, and there are many of them, is cause for concern.

Winter presents the most challenges. Freezing rain, wind chill, deep snow, and cold temperatures, the danger that I might slip and fall on ice, mean that I dress more warmly, carry a cell phone, and use a walking stick. I long ago gave up sledding, but after occasional deep snows, I can still snowshoe.

There is much I like about this season—the often white landscape that sparkles in the sunshine and reflects back its light during long, starry nights, the opening up of the forest that makes it easy to see long distances and spot the occasional animal or bird in the distance, the record of our wildlife residents left in the snow, the birds that flock to our feeders, some of which we only see in winter, the brilliant sunlight pouring from deep blue skies scrubbed clear by cold winds, and, best of all, the surprising wildlife visitors that appear on our back steps, in our yard, out in the field, or in the forest.

My four seasons books are in the American tradition of nature writing that Thomas J. Lyon in his *This Incomparable Land: A Guide to American Nature Writing* calls the "ramble," where "the author goes forth into nature, usually on a short excursion near home, and records the walk as observer-participant." He also describes it as being more or less balanced between natural history information and the author's presence.

As a young woman I had read avidly a wide range of natural history writers and was especially influenced by Edwin Way Teale's four seasons' books—*North with the Spring, Autumn across America, Journey into Summer,* and *Wandering through Winter.* While Teale's books recorded his wide-ranging travels in the United States during each season on a day-by-day basis, I wanted to cover each season day-by-day only on our central Pennsylvania mountaintop. I did write my books in the same order that Teale did and I did record a particular year (spring 1988, autumn 1991, summer 1996, and winter 2001), but because no year could possibly yield as many wildlife sightings as I reported, I inserted happenings from other years that corresponded to the same date in the tradition of the synoptic nature book, a style begun by the English parson Gilbert White in his classic *Natural History of Selborne* back in 1779.

All of the information comes from my nature journals that I have been keeping since we moved to what is now 650 mountaintop acres in west-central Pennsylvania in August of 1971. My hope is that these books and my journals will provide accurate scientific information on what we found here and how it has changed over the decades. Surprisingly, in each case the years I chose ahead of time to record even more minutely for my books turned out to have their own theme— spring is anchored by a red fox family that I watched, autumn by an unsustainable "harvesting" of our neighbor's trees, summer by the birth of our granddaughter, and winter by the slow decline of my elderly father. What surprised me was how tenaciously my father held on to his diminished life and retained interest in and love for his four children, ten grandchildren, and four great-grandchildren. For someone who loved the outdoors as much as I do, he remained amazingly content to stay within his assisted-living facility throughout the long, cold, slippery winter months. I only hope that, should I live as long as he did, I will also embrace life when I can no longer walk my trails every day of the year.

Winter, after all, is the season that most tests our mettle. Getting out day after day, no matter what the weather, can be difficult but ultimately rewarding. If it is difficult for humans, it is sometimes lethal for wildlife. Yet they have many amazing ways to survive winter. So too do the trees and shrubs. Their stories are often as interesting and heroic as those of elderly humans. While this book mostly concentrates on the natural world, it also conveys a little about the winter lives of humans, including those who share my mountaintop.

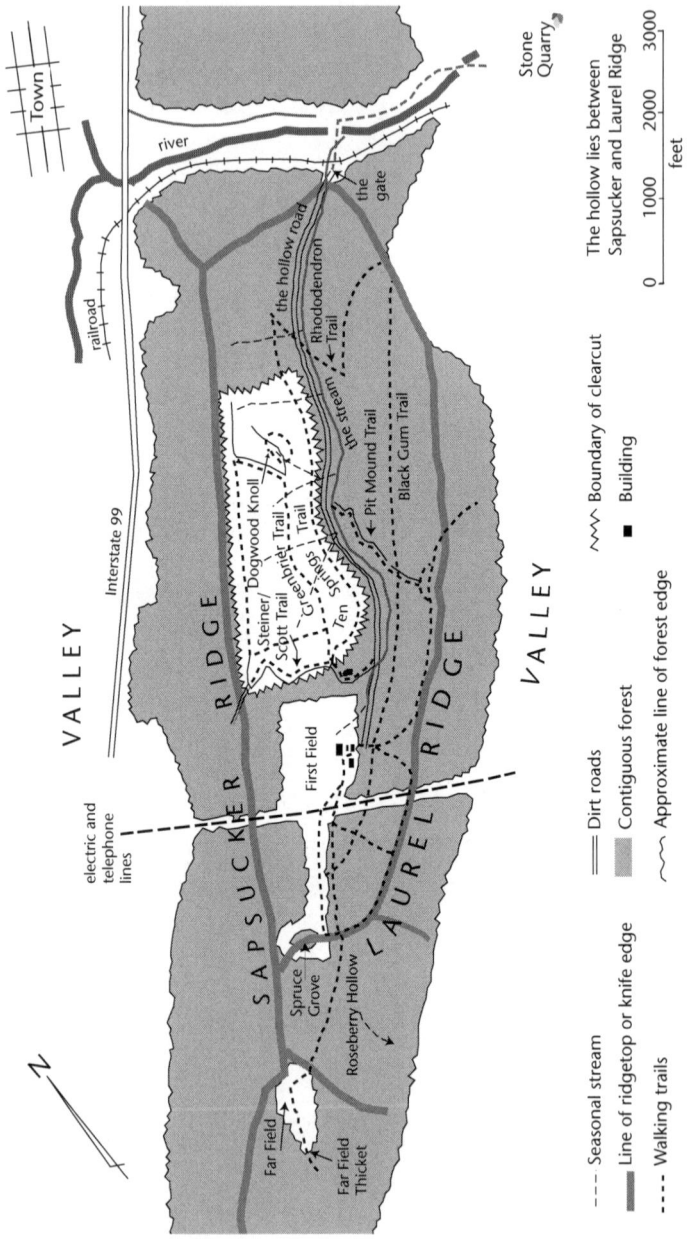

Town

river

railroad

Interstate 99

VALLEY

electric and
telephone
lines

SAPSUCKER RIDGE

Spruce
Grove

Roseberry Hollow

Far Field

Far Field
Thicket

First Field

Dogwood Knoll

Steiner/
Scott Trail

Greenbriar
Trail

Ten
Springs
Trail

the hollow road

Rhododendron
Trail

the stream

Pit Mound Trail

Black Gum Trail

the gate

Stone
Quarry

LAUREL RIDGE

VALLEY

N

The hollow lies between
Sapsucker and Laurel Ridge

0   1000   2000   3000
feet

--- Seasonal stream

Line of ridgetop or knife edge

Walking trails

=== Dirt roads

Contiguous forest

Approximate line of forest edge

∧∧∧ Boundary of clearcut

■ Building

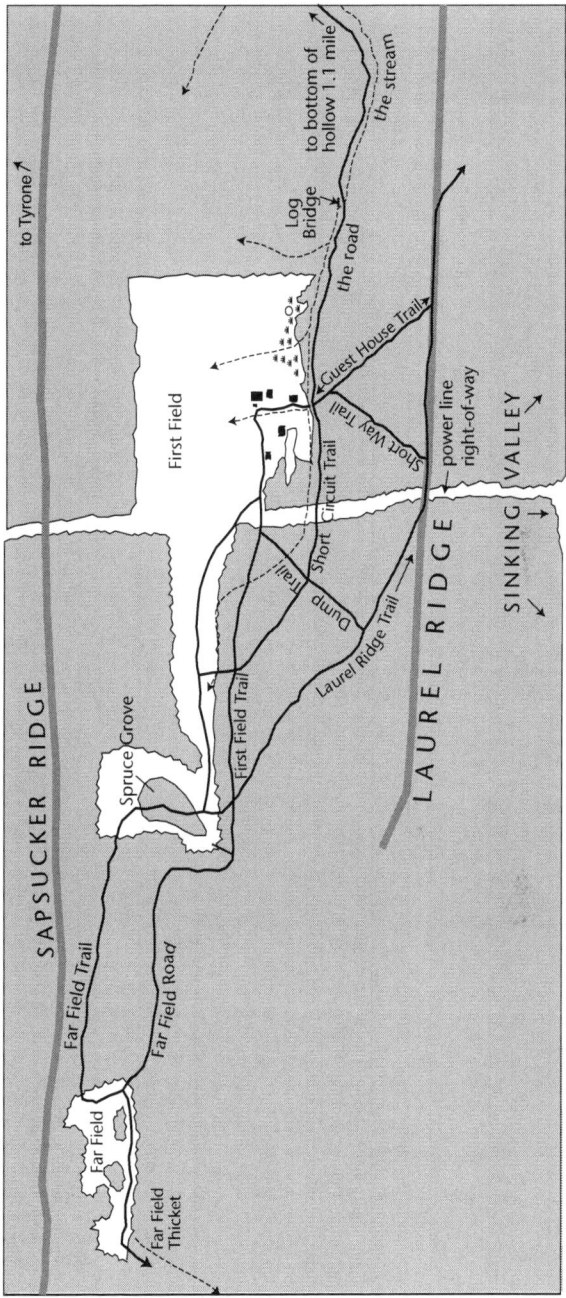

SAPSUCKER RIDGE

to Tyrone

First Field

to bottom of hollow 1.1 mile

the stream

Log Bridge

the road

Guest House Trail

Short Way Trail

Short Circuit Trail

Dump Trail

First Field Trail

Spruce Grove

Laurel Ridge Trail

power line right-of-way

LAUREL RIDGE

SINKING VALLEY

Far Field Trail

Far Field Road

Far Field

Far Field Thicket

N

0   500 feet

——— Road or trail

━━━ Ridgeline

- - - - Seasonal stream (arrow points uphill)

Woods (woods' edge not necessarily field edge)

Appalachian
Winter

In the Ice Forest

I enter the forest of ice
slowly, and on foot.

Trees creak in the slightest breeze.
Small branches break & fall
with a tinkling of bells.

The everywhere green of the mountain laurel
never looked fresher, each leaf
preserved
under glass.

The sun comes out.
A thousand swords leap from their scabbards.

On top of the snow, in every dip
& hollow, windrows of black
birch seeds.

Dave Bonta

December

Go to the winter woods: listen there, look, watch,
and the 'dead months' will give you a subtler secret
than any you have yet found in the forest.

Fiona Macleod, *Where the Forest Murmurs*

DECEMBER 1. He must have tiptoed across our back porch because I never heard him, but as I rehung the bird feeders before dawn, I saw his tracks in the dusting of snow. The night had been bitterly cold and windy, and I had almost told my husband Bruce not to bring in the feeders. What self-respecting bear would be out on such a night? Yet on this first day of December I saw the unmistakable evidence.

When Bruce went for a walk after breakfast, he discovered bear tracks on Greenbrier Trail that headed through Margaret's Woods and across First Field to our driveway.

First the bear had poked around the outside of our guesthouse, and then he had followed our son Dave's well-worn track from the guesthouse to the main house.

At 9 A.M. I took up the bear's trail and tracked it across the backyard and into the open doors of our garage. That's when I decided it had to be a male because he had been checking out Bruce's machinery as any male would. Besides, male bears usually go into hibernation later than females.

After leaving the garage, he circled behind it, crossed our old garden and went into a small patch of woods between the open land and the small powerline right-of-way that bisects our property. From there he had proceeded to the powerline pole, a "mark tree" in biologist lingo. Using a combination of "arched-back rubbing" and tree biting, bears re-mark these places several times a year. Once I even caught a bear in the act.

From the power pole, he more or less followed an old deer trail that winds between the woods and First Field. Because the snow was patchy and thin, I often had a difficult time finding his tracks and frequently had to backtrack. After several hundred feet he reached the Short Circuit Trail, which swings out into First Field to a summer trail Bruce cuts in our old field every year. The bear followed it up into the Norway spruce grove. Perhaps he was resting there.

Because his tracks were larger than my open, gloved hand, I hesitated a few seconds before plunging into the dark, quiet grove. But the grove was bearless and snowless, and I had to skirt the far edge of the spruces until I located his trail emerging from the conifers and heading for the Far Field Road. Anticipating easy tracking on the old woods road, I was dismayed to see that he had climbed down into the sheltered, warm, and fast-melting Roseberry Hollow instead. Even though he paralleled the Far Field Road, the going was difficult as I climbed down and up the steep slope many times, and under, around, and over fallen trees in pursuit of his melting tracks.

In several places I found the melted brown ovals where deer had lain and then leaped to their feet and run off, presumably frightened by the approaching bear. My walking stick helped me move over the landscape, and I used it to dig into the hillside one last time when the bear tracks returned to the Far Field Road near Coyote Bench. From there he stayed on the road, pausing once to dig in the snow beneath a tangle of grapevines in search of fallen fruit.

By then the sun was high and warm, and when I reached

the Far Field most of the snow had melted. Still, I was able to find his fading tracks and follow them as they crossed the field to the old red fox den. For a moment I thought he had wriggled through the small entrance, but the tracks continued to the upper edge of the Far Field and on to Pennyroyal Trail, where they went into the woods beyond the Far Field.

On the melted slope of Second Thicket the tracks disappeared in a brown blanket of dead, hay-scented ferns. After cautiously probing under an uprooted tree with a hole beneath that looked like a perfect place for a bear to hibernate, I gave up the chase. I had been pursuing him vigorously for more than two hours without a rest. Even though I hadn't seen my quarry, tracking the bear had given me a good workout and a little insight into how a bear moves over the landscape.

I rested on Coyote Bench and I watched a red-bellied woodpecker and a northern flicker foraging near the same woodpecker hole in a chestnut oak tree. A winter wren, a high-energy bird in a low-energy season, skipped across the road and into the underbrush where it called as it bounced up and down like a manic jack-in-the-box.

Then two northern flickers—a male and a female—landed on another tree down the slope from where I was sitting. They called to each other and performed a short "wick-a, wick-a" display, spreading their tail feathers and swaying back and forth. Next they flew to a broken-off chestnut oak snag and foraged, she above, he below. When he flew to another tree, he still called and their "wick-a, wick-a's" continued as they foraged on the snag and nearby trees.

According to ornithologists, flickers only give that call when they are defending territory or cementing pair bonds in the spring. At this season, they are supposed to be far south of here, although we often have a flicker or two overwintering if there are enough wild fruits to eat. The literature also says that they probably do not mate for life, but no one is sure.

The pair I watched seemed proof that at least one pair does, and that the "wick-a, wick-a" calls and displays are important for keeping in touch. The fluid natural world and its occupants rarely stay in the little boxes that scientists construct for them, and I frequently observe behavior that either hasn't been recorded by scientists or that contradicts it. Wildlife behavior, like human behavior, varies considerably within the same species.

Once the flickers moved on, so did I. By the time I reached Alan's Bench, in front of the spruce grove, the view was as clear as it ever gets against a cloud-studded, pearl-white and blue-streaked sky. The changing winter sky is one of the season's highlights. Even as I watched, the sun shone on first one blue mountain, then another, and each, in turn, glowed. Although I have admired this view for thirty-two years, I never tire of it.

Scott Russell Sanders, in his book *Hunting for Hope,* says, "Our part in the cosmic story is to gaze back, with comprehension and joy, at the whole of Creation. Our role is to witness and celebrate the beauty of things, the elegance and order in the world, and the Ground of Being that we share with all creatures."

Sanders also talks about simplicity, fidelity, beauty, and other aspects of leading a spiritually rich life. Sequestered up here on our mountain, we interact with hunters, walk the same trails day after day, drive as little as possible, and are without many of the necessities of today's frenzied society—a television, VCR, microwave, and dishwasher, for instance. We try to cooperate as little as possible with a system that tells us to spend more money so that our economy grows. Instead, we try to practice nonconsumptive activities such as reading, writing, walking, and simply being.

My companions are my husband Bruce and our middle son Dave. Both Bruce and Dave are as wedded to this place as I am. Bruce is the problem-solver; the person who runs the machinery and protects our property from the indignities of loggers and developers. As a poet and environmental activist, Dave has chosen to live and work in our guesthouse.

Our lifestyle leaves us plenty of time for family, friends, writing Christmas letters and notes, baking cookies and bread, eating meals made mostly of vegetables and, in season, the fresh food purchased from local farmers in nearby valleys. We try to participate actively in a few local nature and environmental groups and are especially interested in both private and public forest issues. But we also write letters and e-mails and make phone calls about state, national, and even international conservation issues. Most of all, we try to stay informed about what is going on in the wider world. It is enough to keep us busy for a lifetime.

Reading Sanders's book, in which he advocates such a life, I realize that Bruce and I have been doing this for four decades and mostly against the general tide. We raised our three sons here and the other two—Mark and Steve—have retained a love of the natural world and this place even though their career choices have led them far afield. As Sanders advocates, Bruce, Dave, and I have "re-imagine[d] ourselves as inhabitants rather than tourists, cultivating a stronger sense of place, learning about the land, its natural and human history, and the needs of our communities. . . . We [have] . . . learn[ed] to satisfy more of our own needs ourselves . . ."

Recently, Dave has reached out even farther by forming the Friends of Rothrock to defend and monitor nearby Rothrock State Forest. The group actively encourages road closures and the expansion of a state wild area within the forest and counters other threats to this beautiful place from highway projects and gas and oil exploration. Surely a rich country like ours can afford to leave wild places wild. Or must every nickel be wrung from an already impoverished natural world to create the still-elusive goal of more jobs?

DECEMBER 2. Nineteen degrees at dawn and frost blanketed the landscape as the sun rose over Laurel Ridge and shone through heavily frosted windows that sparkled in the light.

I stepped outside in the cold to listen to winter's silence, broken only by the quiet chitter of birds at our feeders. Even the Carolina wren and song sparrow, winter's persistent singers, were struck dumb by the cold. Usually we don't see frosted windows until January and suddenly we wonder not if we will have winter, after the Indian summer days of November, but how bad it will be.

The ground no longer had the springy feel of warmer days. Instead, it was hard beneath my feet as I took a short walk in the mid-afternoon sunshine. I had so many clothes on I could barely walk—long underwear, sweat pants, two pairs of wool socks, lined boots, lined knit mittens, flannel shirt, dark green hooded sweatshirt, heavy gray winter jacket, and orange hunting vest and hat, the latter because on this Saturday the mountain bristled with eager hunters. Sitting in a sunny, sheltered area on Sapsucker Ridge, bathed in winter sunlight, I didn't feel the cold breeze.

Then I walked on to the Far Field, seeing only deer and crows until I heard birds calling on the hillside. I sat down at the edge of the field and waited for them to come. And come they did—tufted titmice, black-capped chickadees, a brown creeper, hairy and red-bellied woodpeckers, American goldfinches, and white-breasted nuthatches—and for a few minutes I felt as if I had entered the oft-reproduced painting of St. Francis of Assisi.

Leaves skipped along the ground like hopping birds and trees groaned as they swayed in the wind. I wandered home over the sun-washed field as the sun sank below the ridge, grateful, as always, for the hushed silence of a winter sunset.

Much of our winter entertainment is at our bird feeders, both during the day and at night. Every fall, in early November, I hang two bird feeders from our back porch latticework. One is an open, wooden platform feeder that has been batted apart at least three times by black bears and patiently repaired by Bruce. That feeder is almost thirty-seven years old and has great sentimental value to us. The birds also prefer it to our other feeder, which is a sturdy tube reinforced by steel mesh and is, so far, bear- and squirrel-proof. Both feeders are filled with black-oil sunflower seeds.

On the back steps and ground below, I scatter mixed seeds of millet, cracked corn, and sunflower. That setup attracts diverse bird species and some mammals too, especially at night. On a mid-November evening three young raccoons first appeared on the back steps to eat birdseed. Turning on the porch light didn't deter them. Neither did opening the squeaky, inside door, sitting on a chair, and watching them through the screened storm door. When the telephone rang, they looked up briefly. When I talked to them, they also glanced up and sometimes retreated back down a step or two, but they soon returned and looked in at me. Finally, an hour and twenty minutes later, after occasionally staring intently into the darkness, they left.

Young raccoons usually spend the winter in a communal nest with their mother and sometimes other raccoons, as many as twenty-three, in a state of semihibernation, once they have built up a layer of fat to sustain themselves during winter food shortages. Those dens are most often in hollow trees but can also be under tree roots, in rocky crevices, or in remodeled woodchuck, opossum, fox, or skunk dens.

According to Dave, however, those raccoons probably live under the guesthouse, along with a skunk, an opossum, a porcupine, and a groundhog or two. Living above this mammal condominium, he expects to spend the winter listening to the assorted bumps, snarls, screams, and hisses below his bedroom as he has every winter he's been here.

Throughout November the triplets, as we called the young

raccoons, visited most evenings, but we never saw an adult. Often, though, they seemed to be disturbed by something in the forest and would leave. Sometimes I thought I heard a faint sound. Was their mother warning them off?

Then, tonight, a young opossum came to the back porch to eat seed. Unlike the triplets, it barely tolerated the porch light. Any sight or sound of us sent it back down the steps with many backward, hesitant looks. Because young opossums stay with their mothers for only three months, this one was on its own. Although opossums don't hibernate, they are relatively inactive in late autumn and winter, resting in nests of grass and leaves, or so the experts say. They are southern animals and have a difficult time if the temperature dips below nineteen degrees Fahrenheit. Opossums that live through a northern winter usually have deformed ears and have lost the tip of their tails because of frostbite.

This opossum arrived before the triplets' visit at 8:15. After the opossum left, Bruce took me outside to see an unusual alignment of the planets in the southwestern sky, one that won't occur for another hundred years. Unfortunately, Mercury was hidden below the ridge, but a perfect crescent moon shone at the end of a lineup of Mars, Venus, Jupiter, and Saturn. Had we a powerful telescope, we would have also seen Neptune, Uranus, and Pluto. The night was clear, beautiful, and still, and the stars glittered in a black velvet sky.

DECEMBER 3. Today it was even colder—thirteen degrees at dawn and clear. The frost-covered First Field sparkled, each grass blade encased in needled crystals.

Across the field, two large does looked nervously at me, but it is Sunday, day of peace, and they were safe. Still they leaped away instead of remaining boldly in sight as they usually do.

I headed over to the old clear-cut on property we bought after it was logged in 1991 and climbed above Greenbrier

Trail to walk Bird Count Trail. That trail is impassable in summer but remains a refuge for birds and animals in all seasons, especially during winter, when it is warmly lit by the rising sun. I too use it as a refuge and laid down against the old road bank to soak up the silence and sunlight.

Such a place inspires me to contemplate the miraculous creation of this earth and its evolution and to wonder why humans have not honored it more, especially now when we know its story and should be even more humble about altering its intricate complexities. Despite the abuse this mountain has taken over the last couple hundred years, I am immensely grateful that I have been able to live much of my life here among the abundant remnants of a mostly human-altered natural world.

Finally, the cold seeped through my jacket and sweatshirt, and I moved on. Ruffed grouse flushed, deer fled, and the clear, whistled song of a white-throated sparrow broke the silence. Ever since they lost their winter refuge in one of the south-facing hollow streamlets that was shorn of all its trees by the previous landowner, wintering white-throats have been rare here. Recently, though, they have been making a comeback and I am happy to be startled out of my winter reverie by their "poor Sam Peabody, Peabody, Peabody" song.

Frost crystals still sparkled on the bench blind and one crow sat above me, croaking its monotonous warning to wildlife of my presence. Another crow landed below the first one and they cawed back and forth for many minutes. Finally both flew off to Laurel Ridge, still noisily communicating. It sounded as if their cawing had more to do with each other than with my presence since it continued long after they left. Crows, after all, are monogamous and often live in family groups that include offspring as old as five years. Perhaps those crows were mates or siblings or parent and offspring.

In the warmth of midafternoon, a black, fuzzy caterpillar moved slowly across First Field Trail, fully protected by its fur coat. It was the larva of the great or giant leopard moth (*Zeuzera pyrina*), a species introduced from Europe before 1879. Overwintering in larval form by hiding in plant material, the caterpillar stirs on warm days to nibble a few leaves of plantain, dandelion, violets, honeysuckle, maples, or willows, its preferred foods. Although the caterpillar appears to be totally black when ambulatory, when threatened, it rolls into a ball, as it did today, and displays vivid red bands between its body segments. Eventually the caterpillar will metamorphose into a striking white and black moth with a three-inch wingspan and an orange-marked, blue abdomen.

At bedtime a great horned owl hooted, reminding me that soon they will be courting in the depths of midwinter.

DECEMBER 4. White-breasted nuthatches vocalized in the sunny woods beside the powerline right-of-way, heralding a couple of golden-crowned kinglets who flew in low to forage. A pair of downy woodpeckers tapped away too.

On the other side of the right-of-way, a male hairy woodpecker hitched his way up a red maple tree trunk and flaked bark off a horizontal limb, followed by a nuthatch doing the same thing.

A rooster crowed from Sinking Valley, which was peacefully swathed in misty clouds while Logan Valley seethed with sounds from the railroad, the interstate, and other commercial activity. Sinking Valley, lacking even a state highway, remains a bucolic anomaly. The advent of several Amish families a decade ago has helped the peaceful process as they pulled the farms they purchased off the electric and motorized grid. They also practice diversified farming and so we have a gratifying choice of vegetables and fruits throughout the growing season as well as honey, fresh chickens, and eggs. To see a farming community gain, rather than lose, diversity in these days of corporate and macrofarming is encouraging.

Lying still on the bank beside the Far Field Road, I heard a winter wren, crows, pileated woodpeckers, cedar waxwings, and quiet drumming from a downy. This seemed to be a "woodpecker day," as they called loudly in addition to foraging during most of my walk.

Unfortunately, our hunters had a drive on and I was in the midst of it, wondering which direction to go to avoid them. Since this is Monday of the second week of buck season and at least nine have gotten their bucks, I had assumed the mountain would be relatively clear of hunters. Instead, I saw one stationed in his tree stand along Laurel Ridge Trail and another standing in front of the spruce grove, gun poised and ready, as I walked along the Far Field Road.

For years, before we posted our land for hunting by written permission only, hordes of mostly unknown folks dressed in orange ranged over our property, running herds of deer this way and that but not killing many. As a result, our forest suffered from overbrowsing and a loss of biodiversity.

A biologist friend, an authority on deer management, told us that our overpopulation of deer was slowly destroying our forest. Tree seedlings, wildflowers, and shrubs were nipped off as soon as they germinated. The understory was badly degraded, impacting other wildlife, such as ruffed grouse, many songbirds, and wild turkeys, that need such habitat for food and cover.

Our friend recommended having thirty to forty deer taken off our land each year. But according to him, that meant hosting 90 to 120 hunters on our land during hunting season. Already I spent most of the two weeks of buck season, except for Sundays, inside. That many hunters, most of them strangers, would turn our property into a battle zone.

Instead, we decided that we needed a few good hunters who would be safe, careful, and effective. First we included the four hunters who years before had helped us with a vandalism problem. Then we talked to people in the area for suggestions and comments about hunters we were considering.

Our eldest son Steve urged us to accept the local Scott brothers he remembered from high school and whose father, Cloyd, had hunted on our neighbor's property for many years. Our brother-in-law, Bob, a lifelong resident of the area, vouched for a few others. To that core list we added neighbors who had been friendly and helpful over the years. Altogether, we have about twenty hunters, depending on how many of their children join them from year to year. By getting to know the people who hunted on our land, we quickly overcame the safety issue and I've been able to go out every day of hunting season.

They also help us with new trail building and road maintenance, post and patrol our property, and share their bounty with us. Best of all, due in part to their in-depth knowledge of the property, they kill far more deer with far less shooting than when we had our land opened to everyone.

Since we initiated our deer management plan, we have watched our forest develop a thicker understory. For the first time in nearly thirty years, some tree seedlings are not only surviving but thriving. So too are many more species and numbers of wildflowers and shrubs.

DECEMBER 5.   Fifty degrees by midmorning, clear and warm. It looks like winter, it sounds like winter, but it doesn't feel like winter.

Insects floated past in the sunlight. A gray moth flew up from the dried grasses at the Far Field and a spring peeper called from the Sapsucker Ridge woods. As I walked down First Field, an orange sulphur butterfly fluttered ahead of me.

Butterflies in December? Then another orange sulphur spiraled up from the grass below the back porch. This late-flying butterfly *(Colias eurytheme)*, also known as the alfalfa, comes from our own Southwest. When alfalfa cultivation became common after the 1870s, this western species spread rapidly eastward, reaching the Northeast in the 1930s. A yellow and orange butterfly that has multiple broods sometimes as late as November, it has been recorded on December 16 near Philadelphia. Warm spells often encourage its overwintering pupa to emerge prematurely, so it is technically possible to see an orange sulphur long after its caterpillar food plants—alfalfa, vetches, and clovers—are finished.

The day remained a miracle of warmth and slanted light—seventy degrees in the sun—as we sat on the veranda in early afternoon. Winter, in a very short time, can have spring, autumn, and even summer weather. It is the most extreme of seasons at our latitude.

DECEMBER 6. Fierce winds last evening brought the temperature back down to eighteen degrees by dawn. A few chickadees and nuthatches called along Laurel Ridge Trail and, as I sat on Coyote Bench, a hairy woodpecker "peent-peented" loudly. I heard a quiet drumming that I thought was probably a downy woodpecker along with a scolding tufted titmouse. This sheltered, sunny place along the Far Field Road, like Bird Count Trail, is a winter refuge for the birds.

I circled a silent Far Field on Pennyroyal Trail and caught a flash of something that seemed too insubstantial to be a bird. It disappeared into the weeds beside the massive remains of a fallen pasture oak. Still, I pursed my lips and pished and a silent winter wren appeared like a brown spirit, popping in and out of its cover. Then I heard a calling eastern bluebird in the distance. This is a warm but noisy corner of the Far Field where I used to bask in the winter sun before the bypass became an interstate highway and destroyed my peaceful retreat.

So many places have been lost to me because of habitat degradation. Neighbors all around us have cut their land until there is nothing left but the twisted trees and scrubby brush that have no value on the market. Every decade has been worse than the one before it as landowners milk their property for every possible penny and more and more highways are built for people to drive more and more cars and trucks. Will it ever turn around again? Not in my lifetime, I imagine, but at least our island of habitat is safe although degraded by its surroundings.

DECEMBER 7.   On this overcast, windy day, I went for a walk along Black Gum Trail, then down Pit Mound Trail, and back up the road. A few deer scattered, a few crows cawed, but mostly I heard only the wind in the trees. A time to think, instead of observe, when the forest seems empty of life. This morning on the radio I listened to a story about the latest space shuttle launch, and I wondered why we continue to spend huge sums of money on space programs to discover what is out there, when we don't know what we have here on earth. We refuse to spend even a fraction of the money we have lavished on space to find out, but we do know that so far there is no sentient life on the planets in our own solar system. Yet we seem to be far more interested in possible space creatures than in the plants and creatures we share this planet with, if interest is equated with how much money we are willing to part with.

We don't even have a working list of all the forms of life on earth. Taxonomists are scarce because there are few jobs for them, and those jobs that are available pay poorly. Our state and national legislatures refuse to give more than a pittance to natural history research despite lobbying by interested citizens year after year. It is appalling, frightening even, how little we know about the workings of our planet, which we depend upon for sustenance.

I assume we continue to throw millions of dollars into space programs because we want to find some refuge for at least a few humans once we have totally destroyed life on earth. After all, it has always been our habit to foul our own nests and then move on to another pristine place on earth.

Now that there are no more such places to escape to, humans are hopeful that somewhere, in the vastness of space, another productive globe spins as beautifully as ours does and that our scientists, who have attained godlike stature, will figure out a way to get us there before it is too late.

DECEMBER 8. Lightly snowing at dawn, with two inches on the ground. The first mourning dove appeared in the feeder area, sitting serenely on the white ash tree overlooking the back porch, her pink breast rosy in the dim light.

There were on-and-off snow showers during my walk, a snow so fine that I often was not aware it was falling. I moved quietly through the muffling snow, walking stick in hand. A pileated woodpecker called in the distance, and gray squirrels were abroad, rushing up and down trees, chasing each other and gathering acorns.

I sat on Turtle Bench looking through my hand lens at many perfect, six-pointed star snowflakes that landed on my dark green sweat pants and slowly melted. As I watched, I remembered what W. A. Bentley, the Snowflake Man, had written over a century ago about each snowflake: "It is an idea that has dropped from the sky, a bit of beauty incomparable, that if lost at that moment is lost forever to the world."

W. A. Bentley was born in February 1865 on the family farm in Vermont halfway between Burlington and Mount Mansfield. His father was a hard-working, dour New England farmer and his mother a former schoolteacher. He had an older brother, Charles, who was much like his father.

Wilson Alwyn was a little different, showing an early interest in science and nature in addition to the basics that his mother taught both of her sons. His mother must have been unusual for those days; she recognized her younger son's brilliance and added Latin to his school curriculum. She also scrimped and saved to buy him a telescope in his early teen years, which he used first to look at the stars. Then he took it apart and improvised a microscope for peering at pollen, insects, and eventually snowflakes.

In an area that often has a ten-foot snow cover for several months of the year, looking at snowflakes would seem to have been a logical pursuit for a scientifically-minded person. But the neighbors didn't understand, his father and brother were unsympathetic, and, as Bentley said to a friend in later years, "Some folks call me crazy. They want to know what good it does to get all those pictures of just snow. . . . I do know what I like. And I think I see as much around here to enjoy, right here on my farm, as those who call me a fool. Most of them have never seen a darned thing."

One person, though, did not consider him a fool when he asked for a camera and microscope. His mother, visionary that she was, persuaded her husband to spend one hundred dollars they didn't have to buy Wilson what he wanted. It took a whole year of saving to accumulate that much cash but she did it, and Wilson was on his way. Not to fame and fortune, however, for even when his snowflake pictures became world-renowned and his articles about them appeared in prestigious magazines, he still spent more money on equipment than he ever received in lecture fees and from writing.

When his father died, he and his brother divided up living quarters in the huge old farmhouse, Charles on one side with his wife and family and bachelor Wilson on the other, and they jointly farmed the acres. But whenever it snowed, Wilson bundled up and headed out to his "snowflake shed," an open-ended, wooden structure he had built to house his camera and microscope. He caught snowflakes on a blackboard that he called "this tabletop from fairyland." Working quickly, he would look at each snowflake under his microscope in the unheated shed. If it was a good specimen, he photographed it through the microscope. In that way, using the equipment his family had bought him, he amassed an enormous collection of snowflake photographs.

Later, when his fame spread, the legend grew that no two snowflakes were ever identical. Bentley never said that. He thought that, given the laws of nature, there would be duplicate snowflakes.

However, with all the diligence in the world, Bentley could photograph only so many snowflakes per storm. Some winters produced two or three good chances for studying snowflakes. Other winters had as many as fifteen ideal storms. He maintained that beautiful snowflakes came from storms sweeping over wide areas, usually from the western or northern portions of storm areas, and that the best temperature was near zero degrees Fahrenheit. In one good storm, he could photograph up to fifty flakes, but he set his own record on February 14, 1928, when he photographed one hundred individual flakes.

Although Bentley was "discovered" by a University of Vermont professor, George H. Perkins, in 1898, and began publishing his research in *Harper's*, *National Geographic*, *Monthly Weather Review*, and other well-known magazines of the day, it was not until 1924 that the official scientific community recognized him with a small research grant given by the American Meteorological Society. In 1931 his book *Snow Crystals* appeared. It is still a beautiful book to leaf through. Large photographs adorn every page with no text except for an introduction at the beginning.

Bentley had only three weeks to enjoy his success. A week before Christmas he caught a cold and stubbornly retreated into his side of the farmhouse to treat it. Refusing all aid until it was too late, he died of pneumonia on December 23, 1931.

An independent-minded Vermont farmer, he just happened to believe that "There is no surer road to fairyland than that which leads to the observation of snow forms. To such a student the winter storm is no longer a gloomy phenomenon to be dreaded. Even a blizzard becomes a source of keenest enjoyment and satisfaction, as it brings to him, from the dark, surging ocean of clouds, forms that thrill his eager soul with pleasure."

DECEMBER 9. A couple dozen dark-eyed juncos flew into the feeders this morning and then, suddenly, they were gone. There, sitting on the ground below the back steps, was a "blue darter," a female sharp-shinned hawk. Her nickname was reinforced by the blue hue of her gray back in the early morning light. She flew to a weed head, landed, flared her tail, and then took off without a meal.

This is the last day of buck season and the first of antlerless deer season, so shots were coming from every direction. I hope our hunters will fill their tags and that will be the end of it, except for a few muzzleloaders after Christmas. It was a day of slaughter as far as our deer were concerned—three more bucks and twelve does were shot. But our hunters were happy. So too was my friend, Chris Bolgiano, an Appalachian forest writer and expert on mountain lions who lives with her husband in rural Virginia next to the George Washington National Forest.

As she told me, "Ralph got his first deer at age fifty-three after teaching himself to hunt in his forties. . . . This is more meat than we have eaten in decades, but boy is it good. We have had a complex of emotions about it, but ultimately feel we are confirming the terrible beauty of life which of necessity consumes itself." An excellent way to put it for nonhunters such as we, and I can't deny our delight when our hunters appear with choice roasts and other parts of the deer they give us. We eke it out for months and feel a tie to our land through eating the bodies of our "wild cattle."

On the other hand, to me wildlife is much more interesting alive than dead, and since I often see signs of caring among animals, I could no sooner kill an animal than a person. Years ago, before we posted our land, several hunters organized a drive right below our house. I watched horrified from our bow window as they fired over and over at the deer that had been spending the day bedded down in the field grasses. Instead of running, the remaining deer stood looking down in bewilderment at their dead and dying companions. Deer obviously live in distinct families or communities since the same deer seem to hang out together—sometimes a doe and her fawns from a couple years, other times a couple of bucks, still others a mixture of bucks, does, and fawns. Every year these communities are disrupted. Yet, because of mismanagement and the loss of natural predators, deer must be hunted for the overall health of the forest, but it is always a sad time for me.

DECEMBER 10. Sixteen degrees at dawn and mostly overcast with a scarlet sunrise warning of bad weather to come. Gray squirrels scampered busily over the lawn in search of black walnuts that still lay in piles. We used to harvest them, but then we discovered that if we left them for the squirrels, they rarely bothered our feeders. Squirrels always prefer wild foods, and if the acorn crop is good, they only supplement their diet with the walnuts.

I took a short walk to Turtle Bench to sit among the big trees, red and white oaks over two hundred years old. Missing even one day outside is difficult for me and like photographer Luiz Claudio Marigo, "I'm a nature addict. I am filled with wonder and a sense of fulfillment in the forest. I need to be here, alone, in contact with different forms of life, away from ordinary human concerns and egos."

I sat on the bench for as long as I could stand the cold, peering down into the Magic Place. A deer moved back and forth like a wraith. Was it the only survivor of its particular community?

Two common ravens flew past in tandem, swooping up and down in perfect synchrony, like a pair of ballet dancers, as I headed home.

Reluctant to go inside, I sat on the back porch close to the feeders in the warm sun. Tufted titmice and white-breasted nuthatches swooped in and out of the feeders; dozens of dark-eyed juncos fluttered in nearby trees and shrubbery as they waited for me to leave, while mourning doves stayed even farther away and higher in the trees. They and the house finches called from a safe distance.

Sitting on the back porch in the winter sun is a favorite pastime, especially when I want a closer connection to living, wild creatures. Years ago, I even handfed black-capped chickadees and tufted titmice, but lately I have been content to merely watch them.

Today is the anniversary of Bruce's retirement from the Penn State Library. He had two parties in his honor. One was library-wide and the other was given by his coworkers. Many nice things were said about his kindness, mentoring skills, and knowledge of the library collection, and hundreds showed up for the library-wide party. Although I was pleased to hear the many accolades for him, I was depressed at the thought of retirement. It signaled that we were moving into the last stage of life, that we will, in not too many years, be leaving this earth to make room for others. Where has the time gone? What have we done with our lives? How much more can we do?

These questions are especially poignant as I watch my father retreat into helpless old age. For a once vital man to sit inside an assisted-living facility, dependent on us to visit, take him to doctors and on outings, rather than outside in his beloved yard, working in his garden, building stone walls, or merely observing the natural world around him must be difficult. Yet he continues to treat his setbacks with grace. He is especially happy this year because our youngest son Mark, his wife Luz, and our granddaughter Eva are house-sitting his country home while Mark writes his dissertation. They visit him often and Eva, who is five years old, is touchingly fond of her Poppop. She is his oldest great-grandchild and eagerly shares her life with him, brimming with stories from her preschool nearby. She also spreads joy throughout his hall, going from room to room to visit and talk with the other residents. Dad calls her "the little girl" and there is no mistaking whom he is talking about. Remembering names is one of the ways he is failing, but his love for children, especially Eva, shines through.

DECEMBER 11. The feeders provided entertainment from morning until night. Early in the morning I recorded six mourning doves, six house finches, and many dark-eyed juncos plus a white-breasted nuthatch and a tufted titmouse. Late in the morning, a rare treasure came to the wooden feeder—a female purple finch eating sunflower seeds as if she were starved. She completely ignored several house finches and the white-breasted nuthatch that flew in, and they ignored her.

In the early afternoon, a black-capped chickadee sat calmly on the edge of the wooden feeder, cracking a sunflower seed between its toes. While watching the feeder birds from the porch door window, I also spotted a pair of common ravens flying past above Laurel Ridge and several gray squirrels working busily and chasing off other squirrels in the woods.

Best of all, though, was our night watching. The triplets and opossum have been visiting most evenings so we continue to check on them by turning on the back porch light. Mark, Luz, and Eva were visiting for the evening and it was Mark who turned on the light. There sat one flying squirrel eating seed on the porch and another on the ever-popular wooden feeder. Eva was enchanted by the small, gray-and-white creatures with huge, dark eyes, and so were we.

Flying squirrels are a mystery to most people. Even mammalogists have a difficult time figuring out their life history. Being nocturnal creatures, these squirrels are active when we are not. They venture from their tree dens about thirty minutes after sunset and, during warmer months, remain active throughout the night. In winter, bursts of activity occur intermittently, with peaks after sunset, near midnight, and shortly before dawn.

Strictly speaking, flying squirrels do not fly, they glide, spreading flaps of furred skin called patagia that extend from the forelimbs to the hind limbs. Using their flattened tails to steer, they glide on average twenty-six to forty feet after launching themselves from a tree. Eventually, the ones we watched climbed to the top of the porch, spread their patagia, and volplaned into the night. No doubt they had a nest in a nearby tree cavity, and our bird feeding station was one of several stops during their nocturnal search for food.

Although there are forty-three species of flying squirrels in the world, only two live in North America—the northern flying squirrel *(Glaucomys sabrinus)* and the southern flying squirrel *(Glaucomys volans)*. The larger, ten- to fifteen-inch, rust brown, heavily furred, northern flying squirrel lives mostly in the conifer forests of the northern and western United States and Canada. The smaller, eight- to ten-inch, grayish, thinly furred, southern flying squirrel inhabits the pine-hardwood forests from southern Quebec and the eastern United States south through Mexico to Honduras. The two species often overlap in the mixed woodlands of the Appalachians and the Great Lakes regions, but most of Pennsylvania, including our forest, is exclusively southern flying squirrel territory.

Southern flying squirrels, also called "white-furred flying squirrels," "glider squirrels" and, the name I like the best, "fairy diddles," are more often heard than seen. They make sounds like the chirping of birds, high-pitched "tseet-tseets," "chuck-chuck" notes, and soft sneezelike calls as they glide through the forests.

They follow specific well-established gliding routes between sheltering sites, even if those routes are not always the most direct or easiest. Memory, not sight, leads them from place to place. One study, near Frostburg, Maryland, found that they liked to forage in the interiors of mature forests with thick understories and especially favored large red and chestnut oaks, red and sugar maples, and American beech trees (all of which we have in abundance in our forest) as nesting sites.

Southern flying squirrels are primarily nut- and seed-eaters, preferring hickory nuts above all other foods, closely followed by acorns and beechnuts. That is why they mostly inhabit mature, nut-producing, deciduous forests. Like other squirrel species, southern flying squirrels collect and store nuts for the winter, caching them in nest cavities and tree holes. Because they don't excavate their own cavities and instead depend on the abandoned nest holes of woodpeckers, especially those made by the downy woodpecker, southern flying squirrels live where woodpeckers flourish. In addition to a primary nest fifteen to twenty feet above ground, they also need several secondary nests for eating, defecating, and retreating to when their primary nest is threatened.

Their winter nests are especially important for keeping warm and are carefully lined with shredded grapevine bark, leaves, and grasses. Southern flying squirrels are known for nesting together in winter aggregations as high as fifty individuals, but usually an average of five to seven huddle in the same nest, saving energy by sharing their body heat. Although southern flying squirrels have low metabolisms during the winter and spend much of their time sleeping, they do not lower their body temperatures or go into torpor.

Late in February the peaceful aggregations of male and female southern flying squirrels are temporarily disrupted by the onset of breeding. After considerable commotion, squabbling, and scolding, southern flying squirrels mate. Then the nesting aggregations settle down again until the females reach the end of their forty-day gestation period.

A week or two before her young are due, a female southern

flying squirrel seeks out her own nest cavity, lines it with soft materials and, in Pennsylvania, gives birth to two to four young sometime in late April. The youngsters will stay with her until she has a second litter late in summer.

The young are born blind, deaf, pink-skinned, and hairless. The female broods them by crouching over the youngsters with her spread patagia. Some females will abandon their young during the first week if stressed or threatened by predators, but after that period, they will fiercely defend their young.

At one month old, the young are fully furred. By seven weeks, they are nearly adult-size. A week later, they are almost completely weaned, and are busy practicing gliding and foraging in trees. At that time, southern flying squirrels eat a wide variety of insects, especially June bugs, wood-boring beetles, and grasshoppers. They also like fruits, such as Juneberries, and seeds, specializing in the seeds inside the pits of wild black cherries of which we have thousands in late summer. The most carnivorous of squirrel species, they prey on birds, eggs, and nestlings, and even eat carrion. In turn, they are preyed upon by great horned and barred owls, black rat snakes, weasels, bobcats, coyotes, foxes, skunks, red-tailed hawks, raccoons, and domestic cats.

The best chance to see a flying squirrel is at a bird feeder. One researcher watched her feeders for years and saw as many as twenty-eight at a time. But the modern record belongs to a Lebanon, Pennsylvania, homeowner who, in 1949, claimed one hundred flying squirrels sailed in to take peanuts. Tonight we felt privileged to see two. We only hope they steer clear of our feeders when the triplets visit.

DECEMBER 12. The day was fiercely windy with falling temperatures, and what snow was left had frozen. But the first American goldfinch appeared at the feeder, closely followed by a second one. Two blue jays also swooped into the wooden feeder and then one flew to the roof outside my bedroom, where it drank thawed water from the storm drain before flying off.

Today, Sunday, is blessedly quiet and I walked over to the sheltered Greenbrier Trail, where a large flock of white-throated sparrows twittered in the underbrush.

Coming back up Ten Springs Trail, I encountered a porcupine moseying along. When porkies are active, they are very active, bustling about with the energy of a roly-poly human, their rear ends swaying back and forth as they move around, their long-clawed front "hands" moving in rhythm with the rest of their bodies. As I stood watching, the porcupine ranged back and forth on either side of a fallen log, its back of quills splayed like a halo. After licking the log, the porcupine finally settled in a hollow area beneath it that was swathed in vines. I couldn't decide if it knew I was there or not, but its final resting spot beneath the log was impregnable.

At sunset I set out in the lessening wind and twenty-five-degree cold for Alan's Bench just as the last rays of sun were shining on Tussey and Nittany mountains, casting a rosy, shadowed hue on the mountains while small, pink clouds scudded above. Then dark-eyed juncos flew one by one into the shelter of the Norway spruces, filling up the tree branches like humans in an apartment complex. My pishing brought several of them to within a couple feet of me, their scolding noises sounding like the clicking of castanets. Then they retreated back into the spruces for the night.

Later, driving back up our road, I spotted a red phase eastern screech-owl sitting in a tree. Even when we stopped and shone a flashlight on it, it never blinked or moved.

DECEMBER 13. A hunter-free woods! What a joy to go out without blaze orange on for a couple of weeks. But underfoot the snow crunched like hyperpopcorn and I couldn't hear anything unless I paused or sat still. By 10:30 A.M. it was already clouding over for forecasted snow/sleet. Bruce put chains on the back tires of our Pathfinder and took Dave along to throw branches off the road and spread ashes at the bottom of the mountain near our gate, which is a sheet of ice. Just in time too, because an ice pellet storm later in the day left three inches on the ground and was followed by warming temperatures and a light rain.

At the feeder area after lunch a red-bellied woodpecker sat on the ground eating seed. I well remember when that species was a rare visitor to our mountain back in the 1970s, and now it is so common that we rarely give it a second glance.

According to the late Merrill Wood, author of *Birds of Central Pennsylvania*, red-bellied woodpeckers had been rare visitors to our area before 1960. After that date, they began breeding locally but were still uncommon.

They appeared here just at the time the effects of the 1980–81 gypsy moth invasion (our first and only outbreak so far) had led to many dead trees. These trees had attracted more woodpeckers and we noticed increased numbers of our usual species—the downies, hairies, pileateds, and northern flickers. Undoubtedly the red-bellieds had also found an abundant source of food in their favorite northern environment, an oak woods, and so had moved in for good. Experts also believe that their extension north is due to maturing forests and more backyard bird feeders because they eat seeds and mast as well as fruit and insects. Red-bellieds, like northern cardinals, are principally a southern species, but they gradually extended their range northward during the twentieth century. In the 1930s our state ornithologist George Miksch Sutton reported that red-bellieds were fairly common in all of Pennsylvania's southern tier of counties and that they should be looked for in the central counties. Since the 1980s they have been expanding in Pennsylvania at 11 percent a year. Today they live even in southern New England, North Dakota, and along the southern Great Lakes' states.

Red-bellied woodpeckers are large birds—9½ inches—with long, needlelike bills and barred, black-and-white backs, which give them two of their nicknames, "zebra" and "ladder-back." The males sport bright, glossy scarlet on the tops of their heads and down the backs of their necks while the females have only the red on their necks and the immatures are brown-headed. They all have white rumps and plain grayish breasts with a hint of red, which gives them their common name.

Red-bellied woodpeckers emit distinctive calls, the most common being a squirrel-like cry, "chiv-chiv." The other nicknames for these birds, "chiv," "cham-chack," "jam-jack," "ramshack," and "sham-shack," are all attempts to describe this call.

Once red-bellieds move into an area they become permanent residents, preferring, in Pennsylvania, to excavate nests in oak trees high above the ground. The females lay an average of five eggs and both parents share in incubating the eggs for fourteen days and then in the feeding and care of the young.

Because red-bellieds consume more vegetable than animal matter, they do well even during hard winters, eating berries, corn, and acorns that they have stored away in trees and fence posts much like their close western relatives, the acorn woodpeckers. They also eat a wide variety of insects and are great fruit-lovers, relishing both wild and domestic fruits. In the South, they are known for their fondness for oranges. They even prey on other birds and have been observed eating the eggs and nestlings of American redstarts, Carolina chickadees, hairy woodpeckers, house wrens, and others. These opportunistic birds will feed at sap holes made by yellow-bellied sapsuckers and have been documented eating tree frogs and small fish, the latter probably scavenged from lakeshores.

Since they are not shy birds, their presence on the mountain adds a noticeable bright note to the winter scene and especially to our bird feeders. And their sharp bills keep even the larger blue jays away.

DECEMBER 14. I was awakened at 5:15 this morning by a pair of great horned owls calling back and forth—essence of winter nights. They went on until 5:30, but Bruce claimed that they had started shortly after 4:00 A.M.

Overnight it froze hard and knowing that I could not walk safely over the icy snow, Bruce drove around First Field up to the spruce grove and then to the Far Field with the tractor. That broke up the ice and made a nice trail for me to walk late in the morning. Dave reported fresh black bear tracks on both the Greenbrier and Black Gum trails. One of our hunters told us that deer gut piles have attracted bears and he is right.

The triplets were at the back porch step at dark. I saw the first sign of aggression between them when two of them pushed the third one off the steps. The victim didn't protest; it merely fed quietly on the ground below its siblings. Perhaps, the two aggressors were male and the victim a female. Studies show that more males than females survive as juveniles. Even though most aggression in raccoons occurs during mating season when the males are defending their females from other males, observers have occasionally reported it at artificial feeding sites such as our back porch.

DECEMBER 15. After a day spent inside visiting with my father, near sunset Eva insisted on going for a walk with me. Up First Field we went, I on the frozen, tractor-tire trail Bruce had made, Eva purposely slipping and sliding, crawling and falling off the track. She was thrilled by the ice and cold.

Remembering a previous visit, in the early autumn, she asked what had happened to all the insects. I told her that many had left eggs to hatch into new insects next spring.

"What about the praying mantis?" she asked.

She had been particularly impressed with the several that I had showed her. After a good deal of searching, I found the hardened, beige froth of a mantis egg case clinging to a goldenrod stalk. As we stood looking at it, a meadow vole zipped in and out of its icy hole in the field, another source of great excitement for the little girl.

I expected to turn around halfway up First Field, but Eva insisted on making it to the Norway spruce grove where we had spent many happy hours the spring before sitting beneath an active American crow nest and listening to the nestlings as they greeted their returning parents laden with food.

By the time we reached the top of the field, it was dusk. After she revisited the old crow nest site, I convinced her to move much faster and stay on the trail as we retraced our steps home. It quickly grew too dark for her to see the icy tread, so I led the way. As we walked downhill we watched first Venus and then one star after another appear. We also heard the barking of dogs in the distance, which she willed to be coyotes.

She was thrilled to be out after dark and walked pluckily onward, not daring to complain since I had warned her earlier that if she insisted on reaching the spruce grove, she would have to get back under her own steam because I could not carry her. But a worried Mark met us at the powerline right-of-way and scooped her up and she filled him with tales of our walk, commonplace enough to me but a heady adventure for her. How wonderful to see the natural world through the eyes of a child. Despite trying to see as much as possible whenever I go walking, I always see more, close at hand, when Eva is with me.

DECEMBER 16. A dull, gray, damp day, with a light mist falling. An unbelievably bad day for our Christmas Bird Count, the worst I can remember in the more than two decades we have participated in this nationwide census. We have had fog and mist and rain before, but never on a ground that is so slippery with ice. Unfortunately, this is the day our local Juniata Valley Audubon Society chose to count all the bird species and numbers in our Culp count circle, which includes our mountain.

Mark was up and out at dawn. Bruce and I were off by 8:30 A.M. after counting large flocks of birds at the feeders, including fifty dark-eyed juncos. Almost immediately the light mist turned into rain. Only a pileated woodpecker and American crow called. On the road to the Far Field we flushed two ruffed grouse. We found no birds at all as we circled the Far Field in the rain on Bruce's tractor tracks.

Then we retraced our steps along the Far Field Road where we heard one tapping downy woodpecker. From there we walked up to the spruce grove where we heard nothing even though the rain had stopped and we were enshrouded in fog. I sat under the spruce trees for protection from the on-and-off mist to write notes while Bruce continued on along the Sapsucker Ridge Trail. But he found not one bird.

Profoundly discouraged, we trekked back to our yard where we saw a couple of golden-crowned kinglets, two black-capped chickadees, and another downy woodpecker. Then a red-tailed hawk screamed in the mist. We probably should have stayed in the yard or, better yet, in the house looking out.

Mark returned at noon. He had scoured Greenbrier Trail and then gone down to the bottom of the mountain where he had walked along the railroad tracks to a neighbor's farm pond, which has always yielded at least one great blue heron. Today it was frozen solid. Finally, he walked up our road, boosting Bruce and my total species of eighteen (including feeder birds) to twenty-eight.

After lunch he headed to the Far Field and the hollows off the mountain in worsening rain. Since I had been thoroughly chilled in the morning, I was not interested in going out again. It seemed a fruitless effort. An all-out thunderstorm and torrential rain brought Mark back in late afternoon with no new species, in fact, with very few sightings of anything at all. But Dave, who had taken a leisurely nap and then carried his coffee out to his front porch in the now balmy midforties temperature, saw a flock of wild turkeys disappearing up Laurel Ridge. He counted four but thought there were many more.

Despite the weather, we had racked up a grand total of twenty-nine species including a hermit thrush, winter wren, brown creeper, and American kestrel. As I remind our hunters every year, they have weeks to get their quarry; we have one day. And we pride ourselves in always doing it on foot.

Checking through my journal, I noted that on this same date in 1997, it was sixty degrees in the warmth of the back porch where I sat most of the morning, soaking up the rays on a very late Indian summer day. Birds flew in to feed, including two female purple finches. I also heard evening grosbeaks and a common raven calling. And, on that day, when Dave sat on his back porch at 4:00 P.M., he saw an eastern phoebe. It broke all phoebe records for the mountain.

DECEMBER 17. Again it is an ice rink outside and I am trapped inside. These are the worse possible winter conditions for me. I don't mind wind, cold, and snow, but I can't go out in the ice. Luckily, it is a Project Feederwatch day so I can spend my time counting birds and species at our feeders. I'm a veteran of this program begun by the Cornell Laboratory of Ornithology back in 1987. Every other week I counted birds two days in a row from November until April and submitted my report on the record sheets they sent me. Then, a couple years ago, I started to submit my records online, which is a lot easier and also enables me to count birds for two days every week if I want to.

Sometimes it seems the birds I get are not terribly interesting, but when I look back over the years, it is amazing what changes have occurred. More blue jays and mourning doves lately, and a year with dozens of American goldfinches. No sign of evening grosbeaks for several years. A few house finches, followed by sixty or more and then few or none because of an eye disease that I never saw in our birds, yet house finches have almost disappeared here. The advent of the red-bellied woodpecker, more and more sharp-shinned hawks, always large numbers of dark-eyed juncos, a few northern cardinals, a waxing and waning and waxing again of American tree sparrows, an overwintering of a song sparrow for several winters in a row, Carolina wrens when I started Project Feeder Watch, a big die-off the winter of '92–'93 and a return at the beginning of this century, and occasionally visits from purple finches and other exciting species, especially during a winter irruption of northern finches—specifically pine siskins and common redpolls.

Today two gray squirrels ate birdseed along with a couple house finches, two American tree sparrows and many dark-eyed juncos. The red-bellied woodpecker also paid an occasional visit and the usual black-capped chickadees, tufted titmice, and white-breasted nuthatches zipped in and out of the wooden feeder. Snowflakes fluttered down even as the sun shone in a blue sky.

In midafternoon the guesthouse porcupine trundled out to the plum tree and Bruce went down to try to photograph it, slipping and sliding over our ice rink yards.

By evening it was nineteen degrees and fiercely windy.

DECEMBER 18.   Still trapped inside but the reward is great—the first ever red-breasted nuthatch at our bird feeder. Was I merely dazzled by his rareness here to think him more attractive than his larger relative—the white-breasted nuthatch? He *was* a male because his cap was black, not gray like that of the female red-breasted nuthatch. Otherwise, he had the coloring of all red-breasted nuthatches—a black eye-stripe below a white stripe and a pleasing, rusty-red breast and belly. His back was the same silver-gray as the white-breasted nuthatch's, but he was at least an inch smaller.

Red-breasted nuthatches prefer coniferous forests for living and breeding and were formerly known as Canada nuthatches because they are common residents in Canadian boreal forests. They particularly like fir and spruce species from the Pacific coast of British Columbia to the Atlantic coast of Nova Scotia and south along the Appalachians in the East and the mountains in the Southwest. However, they will settle for hemlock or pine, especially during the winter when they eat coniferous seeds in addition to their year-round diet of adult and larval insects and spiders.

They also cache food like white-breasted nuthatches do. Usually they cache it under tree bark or in cracks and other interstices. But ornithologists in Montana once watched a male red-breasted nuthatch flip through needle litter on the ground. Carrying a pine seed, he flew to the upper portion of a steep, dirt-cut bank. He probed in the dirt five or six times and he put the seed in the ground as deeply as the length of his bill. Then he picked up a small pebble and tamped it into the hole with a few beak jabs and repeated the same action with a similar-sized second pebble before flying to a nearby pine and foraging on the tree trunk. A tool-using red-breasted nuthatch!

During winters when the coniferous cone crop in boreal forests fails, they sometimes head as far south as the Gulf coast of Louisiana and the deserts of northern New Mexico in search of food. They have even wandered across the Atlantic Ocean to Europe.

The red-breasted nuthatch is the only one of the four North American nuthatch species to engage in these irruptive movements. Usually they occur every two to four years, although at least a few red-breasteds move south of their breeding range every year. My friend Colette, who lives in a nearby town, had a red-breasted nuthatch coming into her feeder every winter for several years.

Here in Pennsylvania, they are nearly always fairly common to common regular migrants in both spring and fall. The fall migration can begin as early as late August and continue through October. In the spring I can usually count on seeing red-breasteds from the second week of April through the second week of May. In the winter both resident and migratory red-breasteds may be solitary, form small flocks, or join mixed-species flocks.

They also seem to join mixed-species flocks when they migrate north. During the spring migration of 1998, I saw dozens of red-breasteds. The best day was the seventh of May when I sat on Dogwood Knoll surrounded by migrating birds. Dozens of yellow-rumped warblers, some only a few feet away, snapped insects from the air, foraged among leaves on the ground, and sat on black locust branches at eye level. Red-breasted nuthatches called from all directions. Many least flycatchers and a few Nashville, blue-winged, and blackburnian warblers joined the yellow-rumps and red-breasteds.

Once the hemlock and white pine forests were cut in Pennsylvania, breeding red-breasted nuthatches were scarce, but the plantations of Norway spruce and red and white pine established in the 1930s started to increase breeding habitat for them by the 1960s. Today, they breed in isolated spruce plantations as far south as York County, west in Beaver County, and even in urban Allegheny and Philadelphia counties, although the Poconos remain their stronghold since they prefer to breed in native black spruce trees.

Red-breasteds form pairs on their breeding grounds, either during the winter, if they don't migrate, or as soon as they return. In addition to courtship flights, when the female usually chases the male in slow motion, courtship-feeding only occurs if the female vocalizes—sings for her supper, so to speak. While engaged in courtship flights, feeding, and subsequent mating, they also begin excavating a nest cavity in dead or partly dead trees. The female does most of the work while the male brings her food and watches out for rivals and/or predators.

Once the nest is finished, the nuthatches smear its entrance with resin from spruce, balsam fir, or pine trees, and continually reapply it outside and inside the nest cavity throughout the incubation and nestling periods. They carry resin globules in on their bills and sometimes use small pieces of bark to apply the resin. This unique behavior trait is probably a deterrent to possible predators and competitors and rarely seems to harm either the parents or the young.

The female incubates between five and eight whitish eggs while the male feeds her both on and off the nest during the twelve to thirteen days it takes the eggs to hatch. Then both parents feed the young exclusively insects for the eighteen to twenty-one days they remain in the nest.

After they fledge the parents continue feeding them for at least two weeks and sometimes longer. Not much is known about either juvenile survivorship or how they disperse after they finish breeding, but a few observers have reported that most youngsters leave their parents' territory and strike out on their own. They may also make up a large portion of the red-breasted nuthatches that migrate south.

Here on our mountain, red-breasted nuthatches are like tiny magicians that pop up when you least expect them and then, just as quickly, disappear. But whenever I hear their nasal, tinhorn call, I know that somewhere there is a red-breasted nuthatch in our forest.

DECEMBER 19. Twenty degrees and snowing, a dry, moderate snow that continued falling throughout the morning, building up to three inches. I looked out at snow-covered gray squirrels chasing in the yard and trying to dig through the ice for black walnuts. At least one gave up and fed quietly on the back step with a blue jay and red-bellied woodpecker, while house finches, nine mourning doves, and mobs of dark-eyed juncos along with an occasional tufted titmouse or white-breasted nuthatch filled in the empty spaces on the porch or wooden feeder.

Meanwhile, I finished wrapping Christmas presents and

planning the rest of my baking projects. Just in time too, because after lunch, heeding Bruce and Dave's urging and assurances that the snow had helped the ice situation, I strapped on my ice-cleated snowshoes and headed into the still-falling snow. For an hour the snow continued as I snowshoed to the Far Field and then back on Sapsucker Ridge Trail to the spruce grove. There, the snow finally stopped after dropping almost four inches. It was easy moving on my snowshoes through a monochromatic world. Although I glimpsed three deer on the powerline right-of-way, it was mostly quiet in the woods.

I was happy to be out after my ice entrapment, but sitting on Alan's Bench, the interstate highway and train noise from the main line at the bottom of our mountain overwhelmed the peace of the forest. Why doesn't humanity realize how greatly our numbers and our technological society are impacting what is left of the natural world? According to a European research group, the World Conservation Union, after conducting the "most comprehensive study of its kind to date, [found that] there is clear evidence that more than 11,000 living species . . . are teetering on the brink of extinction," and that is only a sample of the *known* species. So many have not yet been identified, probably many more than have been.

"We were scared by our own results," Maritta Kock-Weser, director-general of the Union, told the press. ". . . Our world is a result of evolution over 3.5 billion years and we are [doing] away with so much. The magnitude of what we've done is philosophically hard to understand."

Not so hard to understand when I hear an educated person say, "Oh, it's so eighties to be interested in the environment." In other words, it's no longer hip to be an environmentalist. That was just a fad like all the other fads the advertising world has foisted on us. And right now it is time to spend money for Christmas. Not peace on earth, good will to men [and women and the earth we've been given], but spend, spend, spend and keep the good times rolling!

DECEMBER 20. A windy ten degrees at dawn. The January-like weather continues. I snowshoed down the road this morning to escape the wind and found the trees filled with foraging golden-crowned kinglets, black-capped chickadees, and tufted titmice. A few deer tracks meandered along the road, but most of the tracks were made by mice, shrews, or squirrels. Water still flowed through the ice-encrusted stream and the hemlock boughs were white with yesterday's snow, but gusts of wind sent the snow showering down on me. In and out sunlight flickered in the forest. As I snowshoed back up the road, I heard the contentious cawing of crows followed by the high-pitched cries of migrating tundra swans heading for the Chesapeake Bay and other points south along the Atlantic coast.

As always, while showshoeing, I recalled the time, thirty-five years ago, when I became the proud owner of a pair of genuine, handmade snowshoes. During the five years I lived on a Maine farm with my husband and small sons, snowshoes were a necessity, not a luxury, because with all the will in the world, it was impossible to wade through chest-high snow. Then, as now, I needed to be outside exploring the natural world in all seasons, and a pair of snowshoes was my admission ticket to central Maine's snowbound swamps, lakes, and forests.

My snowshoes were specially made for me by a master craftsman, W. E. York of Caratunk, Maine. Because I wanted to use them in the woods, York recommended his most popular design—the modified bear paw, which is oval-shaped with a stubby, squared-off tail.

Even then, factory-made snowshoes were gaining favor, but York sniffed at the newfangled magnesium or plastic frames and nylon webbings coated with neoprene. His snowshoes still had white ash frames, rawhide webbings, and leather bindings.

A winter afternoon in York's woodstove-heated workroom was an education in fine craftsmanship, and we listened respectfully as he explained the step-by-step process he followed every time he made a pair, from his selection of the best ash trees to his construction of the leather bindings.

York was engaged in producing one of North America's oldest forms of transportation. Historians believe that the first device used to extend the human foot for easier snow travel originated in Central Asia about 4000 B.C.

Such devices enabled aboriginal people to move farther north into Scandinavia and Siberia, and also into North America via the Bering Strait land bridge. That bridge became the dividing line between ski and snowshoe users.

The people who moved into northern Asia perfected skis, while in North America the Athapascan Indians in the West and the Algonquins in the East became the greatest snowshoe designers, introducing hundreds of variants depending on weather conditions. Snowshoes allowed them to move quickly over the snow in pursuit of large nomadic game animals such as buffalo, which tended to flounder in deep snow.

Pennsylvania formed the southern boundary of snowshoe use, and both the Delaware and the Susquehannock utilized them. Tradition has it that the town of Snow Shoe in Centre County, formerly Snow Shoe Camp, honored the experience of a party of white hunters who, overtaken by a snowstorm in the area, constructed their own snowshoes and made it into the Bald Eagle settlement. More likely, though, according to historian D. Zeisberger, the town was named for a snowshoe found hanging in a tree in what was once an Indian camp.

The first European people to adopt snowshoes in North America were the French in Canada who, along with their Indian allies, later used them as a tactical aid for making quick raids on British settlements during the French and Indian

Wars. Their successes forced the British to become adept snowshoers. From that point on, the English colonies and, later, the newly formed United States routinely equipped their militias with snowshoes.

Recreational snowshoeing is also first attributed to the French Canadians in Quebec, who started snowshoe clubs more than two hundred years ago. Originally they organized snowshoe races to train military men, but gradually they became civilian organizations with colorful uniforms and sashes identifying their districts.

Most appealing to me were the snowshoe hikes held in New England villages until the early 1930s. Family groups of from thirty to one hundred people participated, and a committee planned a secret trail that would end at a farm where the owners had been previously contacted and paid to provide a hearty meal of homemade soup, sandwiches, biscuits, doughnuts, cider, and coffee.

Afterwards, the snowshoers took a different route back over a hill so they could do some snowshoe sliding. Such outings occurred once a week throughout the winter if conditions were right. This strikes me as a much more healthy sport than sitting on snowmobiles and blasting through the forest as many people do today.

I didn't know any of the history of snowshoes when I bought mine from York. My major concern was whether an unathletic klutz like me could master the sport. I buckled the snowshoes over my high-topped leather boots and started walking. And that was it. It *was* as easy as walking, especially with a light crust on the snow.

Until the blizzards of the early nineties, I persisted with York's snowshoes. Bruce had a pair of his own and he and Dave shared them. But when we had to depend on them to get in and out for several weeks, I decided to buy a pair of the modern, lightweight snowshoes as a fallback. I must admit it was love at first step. Because they were much lighter than my wooden snowshoes, I found myself running downhill on them. Then, too, the ice cleats on the bottom allow me to go out in conditions like we have had the last several days. They are still the best way to see a snow-filled winter woods.

DECEMBER 21. The red-breasted nuthatch was back on and off the feeder all day. A fat meadow vole stuffed its face with the fallen seeds below the steps, popping in and out of the hole leading to its winter burrow that it has conveniently dug in the midst of the seed.

On this first "official" day of winter I startled four deer lying up in Margaret's Woods and a couple of ruffed grouse below Greenbrier Trail, but mostly the woods were still.

The opossum was out on the back step eating seed shortly after dusk. Its coat was pitifully thin. In places its skin showed through, like a balding old man with a few strands of gray hair plastered over his head.

The golden globe of a full moon rose above Laurel Ridge, and we celebrated the winter solstice by snowshoeing over the moonlit field and through the long-shadowed forests. Fallen trees glowed in the white light and the dark silhouettes of deer moved off as we approached.

DECEMBER 22. The skies cleared after an inch of fine snow that fell overnight, so I snowshoed up to Turtle Bench to sit among the old-growth oaks and listen to the trees squeaking and groaning in the wind and cold. Common ravens "cronked" in the distance and black-capped chickadees "dee-deed." As I snowshoed on up to the spruce grove a pileated woodpecker "kik-kiked" repeatedly from the sunlit slopes of Sapsucker Ridge.

In the shelter of the spruce trees, I looked out to see what I thought would be a brief snow shower, falling thickly in the midst of a dimly-shining sun, and another pileated on Laurel Ridge called more quietly, as if subdued by the snow.

I snowshoed along the Far Field Road through gray curtains of falling snow. Then I sat on Coyote Bench and watched the flakes turn to diamonds in the brief sunlight. Later, I made a perilous circle of the Far Field in the snow squall, sliding a couple of times on the ice beneath the dry snow covering the steep slopes, and encountered a flock of black-capped chickadees and golden-crowned kinglets as well as a foraging downy woodpecker. The songbirds were busy despite the snow. Along Sapsucker Ridge Trail, there were still more chickadees and kinglets close-up and calling tufted titmice. This is certainly a big year for kinglets—three separate flocks in the hollow and two on the mountaintop.

I returned to the spruce grove and methodically ranged back and forth in search of owls. A big brown bird suddenly flushed so fast I couldn't decided if it was an owl or a raptor. That unknown was followed by knowns—two American crows scolding loudly over the grove.

Despite the sun and blue sky, snow continued to filter down and I watched a whirlwind of snow spinning over First Field like a sign from God.

I arrived home after noon, having been out three hours and wishing that I could be out even longer. The bird numbers at the feeder started to increase—seven American tree sparrows, a black-capped chickadee, mobs of dark-eyed juncos, house finches, nine mourning doves, the male red-bellied woodpecker, and a Carolina wren. As the wind blew and the temperature plummeted, they fed frantically, as if they could not get the seeds down fast enough.

In midafternoon Dave shouted that a sharp-shinned hawk had nabbed a dark-eyed junco. It plucked and ate it below the back porch. From my study window I had an excellent view of its blue-gray back and red-streaked breast and watched as it scattered junco feathers and pulled out the entrails. Finally, it flew up to a nearby tree, shook its long, banded tail, and flew off. I could almost hear it smacking its bill contentedly. All the while it plucked and ate, not a bird was to be seen. But once it disappeared, it wasn't long before more birds were back, including a male northern cardinal feeding regally among the plebian doves and juncos, tree sparrows and house finches.

Tonight, we went to see Eva in her first-ever Christmas pageant at her preschool. She was one of several angels who managed to sing at least a few songs in between yelling at her Dad in the audience to stop taking pictures. Luckily, most of the other three-, four-, and five-year-olds were equally chaotic, waving to their parents and calling out.

Driving the twenty-five miles home, we saw many houses and yards covered with Christmas lights, some tasteful, others not, but all symbolizing the return of light or, in the case of Christians, the coming of the Light. At the darkest time of the year, this light worship seems to be most appropriate despite what is a profligate use of electricity in some of the more elaborate displays. On the other hand, I'm content with the "moon on the breast of the new fallen snow [that] gave a lustre of midday to objects below . . ."

DECEMBER 23. Five degrees at dawn and clearing into a blue-skied morning. Birds crowded the feeder and porch in the dawn light, eager to stoke up after the bitterly cold and windy night. But when I came back downstairs at 9:15, after doing my back exercises, only three gray squirrels fed calmly on the porch. Not a bird was in sight.

Then I saw that the Killer Extraordinaire had struck again. This time the sharpie plucked its junco victim on the other side of the clump of fallen weed stems from where it had worked yesterday and even closer to our bow window. It saw us watching so it picked up the remains of its kill in its talons and flew away.

I set off on snowshoe close to ten o'clock. It was nine degrees with a light breeze. The old dump area was a maze of mice tracks and tiny holes, meandering in what looked like drunken circles of reverie. Had they had a party last night? Certainly I found no other concentration of mice tracks like that during the rest of my walk. Another more scientific question might be, what attracted them especially to that area?

Woodpeckers—hairies and downies—tapped and called along Laurel Ridge Trail. Sitting on Coyote Bench, though, I heard a calling northern flicker, cedar waxwings, a pileated woodpecker, an American robin—what a great day for a Christmas Bird Count this would have been!

Golden-crowned kinglets and black-capped chickadees foraged along Sapsucker Ridge Trail. My old tracks served as a road way for coyotes, and porcupine tracks paralleled them, threading their way into and out of the edges of the spruce grove.

The sharpie was back after lunch, sitting in the tree below the feeders, preening and shaking its tail. I rushed out to chase it, figuring that one junco a day was enough for it, and it flew toward the front porch. So I went out on the front porch and peered around the side of the house into the ten-foot-high juniper tree in the middle of the old herb garden. The sharpie lurked in its prickly midst, but it flew off when it realized I had spotted it. This must be where it launches its attacks. Despite frequent sharpie visits over the years, this is the first time we have had an unusually proficient and persistent hunter.

The brilliant blue skies of the morning gave way to clouds by afternoon. By evening the sky was slate gray, but the snow on the ground lights up the outdoors even without brilliant skies, and once again we will have a bitter cold, white night.

DECEMBER 24. Eight gray squirrels visited the feeder in the fourteen-degree cold and only the blue jays were bold enough to challenge them.

The sky slowly cleared as Bruce and I walked along Laurel Ridge Trail and the Far Field Road, I still on snowshoes and he in boots. We heard only a few tufted titmice, black-capped chickadees, and golden-crowned kinglets, and then the wind picked up and the high-pitched squeak of trees replaced bird calls. In the depths of the spruce grove, we flushed a big bird but didn't get a good look, although I think that it was a great horned owl.

Bruce was fascinated by the abundance of black birch scales and seeds that blanketed the snow like beige confetti. Marking out one square foot where they were thinly spread, he counted 120. In another square foot thick with them there were 280. And that was on only one layer of snow. My mind boggled as it tried to imagine how many seeds and scales covered our mountain.

Nineteenth-century naturalist and writer Henry David Thoreau, also a counter like Bruce, discovered that "each catkin, one inch long by a quarter of an inch wide, contains about one thousand seeds," according to his "The Dispersion of Seeds," first published in 1993 as part of a compilation of his natural history observations called *Faith in a Seed*. When these catkins are dry, they shake off like chaff in the wind.

The scales of birch cones, he wrote "are three lobed . . . having the exact form of stately birds with outspread wings, especially of hawks sailing steadily over the fields." To me, the scales look more like the fighter planes that fly low, loud, and fast over our home, not surprising since the design of those planes is based on nature's design of birds.

The birch seeds are smaller than the scales and resemble "tiny brown butterflies." We agreed with that assessment as we looked more closely at the brown "dust" skittering across the snow in the slightest breeze.

Thoreau thought that wind was one way birch seed was dispersed. Furthermore, it was not only "blown far through the air, but slides yet farther over snow and ice." Exactly right, as we discovered today.

We also examined the many winged seeds of striped maple trees that have fallen in the wind along the Sapsucker Ridge Trail. The slightest shake of a sapling brought still more down. "Convince me that you have a seed there," Thoreau wrote, "and I am prepared to expect wonders." So were we.

DECEMBER 25. Fourteen degrees and windy at dawn and we are having a partly sunny, white Christmas. To escape most of the wind, Bruce and I took our traditional Christmas walk down our Plummers Hollow road. I was able to go without snowshoes because the snow had packed down in the middle of the road, and I felt strangely light-footed as I walked along.

First a perky Carolina wren darted in and out of hollow logs and debris along the stream just as winter wrens do. I hope such an approach will help get this northward-moving southern species through the bitter cold. Farther downstream, Bruce spotted a winter wren that is aptly acclimated to winter's cold.

In the depths of the hollow a brown creeper foraged on a tree trunk along with the usual contingent of black-capped chickadees, tufted titmice, a downy woodpecker, and golden-crowned kinglets in the hemlock grove. Bruce even glimpsed the "Christmas bird"—a brilliant male northern cardinal—as it flashed across the road.

We again amused ourselves by identifying the potpourri of tree seeds scattered on the road from hemlocks, tulip trees, basswoods, red maples, and, underlying all, the ubiquitous black birches. The wind also brought down one short wild yamroot vine with its distinctive, tan, heart-shaped fruits containing light, flat seeds.

As we rounded the guesthouse curve, there sat a porcupine, looking very unpartridgelike, in a pear tree, specifically the huge old Kiefer pear in the guesthouse backyard, where it was systematically debarking its branches. Soon the porcupine retreated back under the guesthouse where it spends most of its days.

We hurried home so I could make Christmas dinner for

Mark, Luz, Eva, Dave, and my Dad. After opening presents, Eva spent most of the day playing board games with Dad, making up her own, simpler rules. Later, she insisted on going along when we drove him twenty-five miles back to his assisted-living home after supper.

"We had a good time together, didn't we, Poppop?" she said happily as she hugged him goodbye. Dad's grin was all the answer she needed.

The blessed little girl had made his day, and her genuine love for him touched all of us. Almost eighty-eight and five-and-a-half—what a perfect combination.

DECEMBER 26. Every winter birdwatchers hope for an irruption of boreal birds from the northern forests. This "irruption" is an irregular, migratory movement southward of birds that ordinarily live and breed in Canada and Alaska and includes such species as pine and evening grosbeaks, purple finches, red- and white-winged crossbills, pine siskins, common and hoary redpolls, red-breasted nuthatches, snowy owls, northern shrikes, northern goshawks, and rough-legged hawks. This winter all we have seen so far across the state are the red-breasted nuthatches.

The northern songbirds are dependent upon the seeds of conifers and a few hardwood species, mainly alder and birch, and when the seed crop fails, as it does periodically, and bird numbers are high, they are forced to head south in search of food. The same is true of northern birds of prey, including the meat-eating northern shrikes, all of which prey on lemmings, voles, or snowshoe hares, or a combination of all three. When those populations crash, their predators must also migrate south in search of food.

In the thirty-two years we have lived on our mountaintop, we have had occasional visits from northern shrikes, evening grosbeaks, purple finches, white-winged crossbills, and northern goshawks, but we have witnessed the irruption of only three songbird species—pine siskins, common redpolls, and red-breasted nuthatches. The pine siskin irruption years were 1987–88, 1989–90, 1995–96, and 1997–98. Common redpolls appeared in 1993–94, 1995–96, and 1997–98; and red-breasted nuthatches in 1995–96 and 1997–98.

The pine siskin irruption in 1987–88 was one of the largest in living memory. At least ninety-five million siskins appeared at feeders all over North America.

I remember my first sighting of the small, brown-streaked, sharp-billed birds. On October 26, 1987, at the Far Field I heard and then saw a flock of twenty pine siskins eating seeds from a small black birch tree at the edge of the field. Pine siskins *(Carduelis pinus)* both sound and fly like American goldfinches *(Carduelis tristis)*. While the two are closely related, siskins are bolder and they ignored me as I crept close and sat on the ground to watch them. After fifteen minutes they whirled off.

A snowstorm in early November brought them to our feeders for the first time. I looked out almost in disbelief as more than eighty descended, settling on saplings, the ground, back steps, porch floor, and feeders. But they flew off in a few minutes.

Throughout the winter during stormy days the siskins came as a body to the feeders—up to one hundred at a time—and gobbled up sunflower seeds. But because it was a mild winter they spent most of their time in the forest eating black birch seeds, and I spent a lot of time watching them.

On a sunny December day at the Far Field Thicket, sixty of them twittered softly as they fed in a black birch tree. Then most of them flew down to a fallen tree trunk to eat snow. Fifteen of them lined up almost beak to beak. Others ate snow from tree branches and on the ground. After that they returned to eating birch seed.

At the end of February I found eighty siskins running over the snow-free ground at the base of Sapsucker Ridge. The golden wing bars of some of the males were prominent, and in addition to their usual goldfinchlike calls, they also made buzzing sounds that resembled those of blackbirds. Occasionally they swooped up into saplings in response to warning calls, but I never did see what startled them. Mostly, though, they ran over the ground, feeding on fallen tree seeds, sometimes coming within ten feet of where I was standing.

I continued to see them in the woods until the end of April, and then they were gone, off to nest in high altitude coniferous forests. Siskins nest from Alaska east to Newfoundland and as far south in the eastern United States as northern Pennsylvania. Now siskins are one of Pennsylvania's rarest breeding birds, but ornithologists think they were more abundant here before our coniferous forests were cut.

During the 1989–90 siskin irruption, they spent more time at our feeders than in '87–88. Otherwise, they hung out in the hollow with American goldfinches, eating hemlock seeds. For sheer entertainment at the feeders that winter, the pine siskins' antics couldn't be beat. Although there were less of them (forty) than in 1987–88, they totally dominated the feeders whenever they came in, even pushing aside the hordes of house finches. These littlest creatures on the feeders threatened every bird that came too close by running toward them, sharp beaks open, and the males flashed their yellow wing patches like caution lights. House finches, American goldfinches, dark-eyed juncos, American tree sparrows, black-capped chickadees, and even the larger tufted titmice fled.

The pine siskins gobbled pounds of sunflower seeds every day. One of them even blundered inside the tube feeder when I forgot to close the top after refilling it. I found the bird flopping around, trapped by its own greed. When I reached in to lift it out, it seemed to understand what I was doing and did not struggle in my hand as most birds do, nor did its heart beat any faster. Siskins, it seems, are too self-confident and scrappy to be frightened by a mere human.

Common redpolls *(Carduelis flammea)*, in the same genus

as American goldfinches and pine siskins, are similarly un-afraid of humans. They also like birch seeds and breed even farther north than pine siskins—from the southern edge of the Arctic tundra south into coniferous forests from Alaska to Newfoundland. They and their close relatives, hoary redpolls *(Carduelis hornemanni)*, can survive colder temperatures than any other songbirds, probably, in part, because of a special storage pouch in their esophagus, which they fill with food just before night falls. The slow, overnight digestion apparently generates just enough heat to keep them from freezing.

Both species have bright red caps on their foreheads ("redpoll" means "red cap") and black chins, and male common redpolls have pink breasts as well. While hoary redpolls' rumps and breasts are frosty-white, common redpolls are brown-streaked, as are the back and wings of both species.

I have never seen a hoary redpoll and until the '93–94 irruption had seen only an occasional common redpoll in First Field over the previous twenty-two years. So, on January 10, 1994, I could scarcely believe my eyes when, hovering at the edge of the house finch flock, I spotted a nervous common redpoll. Within a couple of hours it was joined by seven others. All through the bitter cold and snow of January and part of February I could expect to see as many as ten, but their last day at the feeder was February 18.

Having waited so long for my first common redpoll irruption, I was amazed to witness another one two years later. Four common redpolls first appeared at the feeders on December 9, 1995, when the thermometer stood at seven degrees. The pine siskins had already been in since November 12, but their numbers were low. Fourteen was the highest count we had all winter. Both species came and went irregularly until the middle of March, with common redpoll numbers peaking on March 9 at thirty-two.

The common redpolls proved to be as feisty as the pine siskins. One afternoon I watched a pugnacious redpoll hold the wooden bird feeder against all comers. In less than a minute it chased each interloper that landed by rushing at it and chittering loudly—a downy woodpecker, tufted titmouse, black-capped chickadee, dark-eyed junco, and American tree sparrow fled in the face of its naked aggression.

That was the same winter I found a red-breasted nuthatch every time I walked down the hollow road. My first sighting occurred on a windy, cold December 21. It was silent in the hollow until I reached the hemlocks. They were filled with a merry band of chickadees eating hemlock seeds. As I stood watching them, I was thrilled to see a red-breasted nuthatch land on a hemlock trunk about six feet from me. After giving me the longest, closest view of its kind I had ever had, it flew to a fallen log spanning the stream and foraged beneath it. It was as quick and energetic as the black-capped chickadees, flitting from tree to tree, up, down, and around at twice the speed of its congener, the white-breasted nuthatch.

The hollow chickadee flock broke up by mid-January when the hemlock seed crop was exhausted, but we had red-breasted nuthatches in the forest until May. While that irruption was weak in numbers, it was rich in species diversity. To look out at my feeders and see both pine siskins and common redpolls, and to walk down our hollow road on a cold, windy day and watch red-breasted nuthatches, was a joy and a privilege that brightened what turned out to be the longest winter of our lives.

The winter of 1997–98 was not only a classic irruption but a superflight in which all the highly irruptive finches, as well as the red-breasted nuthatch, appeared somewhere in Pennsylvania. Although we didn't have the more glamorous species on our mountain—the crossbills, pine grosbeaks, or hoary redpolls—we experienced the greatest winter finch diversity ever, both at our feeders and in the woods. On November 12, twenty evening grosbeaks appeared at our feeders. The following day they were joined by pine siskins.

"It looks like it is going to be a finch winter," I wrote happily in my journal. And indeed it was, even though the evening grosbeaks and pine siskins moved on by the end of the month, briefly dampening my belief in a finch invasion.

But we did have and continued to have huge numbers of American goldfinches, far more than I could ever remember. There were sixty at a time at our feeders, when five to ten had previously been our highest count. And in the woods large flocks coursed back and forth overhead as I took my daily walk.

What were they eating and why were there so many? Our conifer cone crop had failed, but our black birch seed crop was huge. And that was what they were eating. American goldfinches, like their close relatives, are nomadic in winter, going where the food is.

By late December, I was still hoping that those birch seeds would bring in common redpolls and more pine siskins. And on a less than auspicious Christmas Bird Count day in late December my hope became a reality. Bruce and I plodded through light snow for three miles, combing empty ravines and seeing few birds. But late in the morning, as we recrossed the Far Field, we were halted in our tracks by a chorus of bird calls. A grove of black birches, loaded with seed catkins, was also loaded with at least two hundred common redpolls.

We sat down on a log and watched for a long time as they wheeled back and forth over the treetops, then settled down to eat, first on one tree, then another, before the whole flock finally took off.

That was the true beginning of the finch invasion here. Pine siskins also returned, although not in high numbers like the common redpolls. Most days both siskins and redpolls visited the feeders in small numbers (five to ten), along with hordes of American goldfinches, but if I wanted to see all three species in the hundreds, I headed for black birch trees.

I spent many happy winter hours, on sunny days and overcast ones, sitting at the base of a nearby tree and watching them move restlessly from catkin to catkin, chirping continually. One snowy day a mixed flock alternated eating seeds from the catkins with eating snow from tree branches. Other days I encountered common redpoll flocks feeding alone as we had the day of the Christmas Bird Count.

The common redpolls and pine siskins remained on the mountain until mid-April and were joined for several days by more purple finches than usual. After so many irruption years in the nineties, I have waited expectantly winter after winter for another one. But the pine siskins and common redpolls have not returned.

DECEMBER 27. Sixteen degrees at dawn and lightly snowing with a breeze. A male northern cardinal flew into the wooden feeder at 9:15 and continued flying in and out all day. That was unusual behavior for a cardinal because normally they come at dawn and dusk and are very shy. This one was not.

Today is another Project Feederwatch day and I've also counted two song sparrows, eighteen mourning doves, more than fifty dark-eyed juncos, a black-capped chickadee and white-breasted nuthatch, two tufted titmice, the red-bellied woodpecker and red-breasted nuthatch, four blue jays, a couple of American tree sparrows, and eight gray squirrels.

I resumed snowshoeing on the mountain trails since it is still slippery. So slippery, in fact, that Dave can sled anywhere, even through the weeds of the fields. How I envy him and wish I could entrust my body to a sled without damage to my back. I still remember how wonderful it was to whiz down the back slope, as I did when we first moved here. But I must be content with safe and contemplative snowshoeing.

I examined the snowflakes that fell on the black cuffs of my coat through my hand lens and saw several striking versions of the six-pointed star. This dry snow, the same kind that fell in

Bentley's Vermont, is excellent for looking at the shapes of individual flakes because there is no clumping together as in a wetter snowfall.

Snowflakes—or snow crystals, as they are often called—are formed in clouds where the temperatures range from thirty-two to thirty-nine degrees Fahrenheit. These clouds consist of billions of small droplets of water and minute, cooled particles of salt and dust from the earth's surface. Individually, a cooled particle attracts water molecules from the droplets to it. The molecules land on the particle, freeze, and build ice crystals.

Once formed, a crystal does not stay fixed in shape. The temperature of the atmosphere and the amount of water vapor in it determines the way a snow crystal grows. As it falls slowly toward earth, buffeted by winds and updrafts, moving continuously through a wide range of temperature changes, the snow crystal alters its shape constantly.

After it reaches the earth, it can be classified into several general types of snow crystals by an observer with only a hand lens. In addition to the star-shaped or stellar crystals, there are hexagonal flat-plate structures; column crystals; and thin, needlelike crystals.

These simple crystals, in turn, may change into more complex forms as they fall. The columns may be capped on either end by the flat-plate structures, forming capped column crystals, or pieces may break off the stellar crystals and reform as spatial dendrite or branching crystals. Finally, there are the so-called irregular crystals that have been knocked about so badly in the atmosphere that they no longer have a discernible shape.

But is each snowflake truly unique? Bentley didn't think so. Other scientists, though, have been working on the question.

Back in 1973, Charles and Nancy Knight wrote an article about snow crystals in *Scientific American* magazine and answered the question mathematically. Since the typical snow crystal weighs a mere millionth of a gram and consists of one billion billion molecules of water, "considering the huge variety of ways that number of molecules can be arranged, it may very well be that there have never been two identical snow crystals," they concluded. Scientists estimate that one hundred billion trillion trillion snow crystals have fallen on the earth.

While such incredible numbers sound convincing to a mathematical moron like me, scientist James Langer at the Institute for Theoretical Physics has been reexamining the question of why snowflakes form as they do, using his expertise in physics, math, and metallurgy to simulate snowflake formation on a computer. He believes that snow crystals are barely stable elements, so as each one drifts slowly down to earth encountering a slightly different set of temperatures and humidities, it reacts to every change in conditions by slightly altering its shape, and because each snowflake takes a different path from every other flake, it is probable that every snowflake is unique.

I sat on Coyote Bench in sunshine that made the still-falling flakes sparkle. There were no signs of hunters today, even though muzzleloader season began the day after Christmas. We did see and hear some evidence yesterday that they harvested more deer, but maybe everyone is back to work today.

Gray squirrels were out and about. One leaped from tree to tree via low, thin branches and finally ran up a large tree. Instantly, loud scolding erupted and it was chased down that tree and over several smaller ones until it was back in its original area. Then the chaser retreated to its large tree. Could it be a preliminary to mating, which is coming up soon?

DECEMBER 28. Ten degrees at dawn and windy. We had a dusting of snow overnight, but it cleared to a mostly sunny, gusty day, the snow blowing in the wind and white as the gypsum sand we saw at White Sands National Monument in New Mexico.

I walked down the hollow road to escape most of the wind. A downy woodpecker, a large flock of black-capped chickadees—one even singing "fee-bee" in the sunlight—and golden-crowned kinglets moved slowly down the sunny, Sapsucker Ridge side of the mountains into the still-shady depths of the hollow.

If I were to choose my favorite winter bird, it would be the golden-crowned kinglet. Bold and cheerful as chickadees, golden-crowned kinglets are smaller, more elfin and full of fluttery grace, the Tinkerbelles of the bird world. Almost entirely insectivorous, they have no use for bird feeders. They make their living by gleaning insects, their larvae and eggs, as well as mites and spiders, from tree buds and conifer needles and under leaves and bark, often hanging upside down to do so.

I remember one bitterly cold day in early January, when the silence was broken by the high-pitched "tsee-tsee-tsees" of golden-crowned kinglets. As I stood quietly watching, the kinglets flew within a couple feet of me and foraged low to the ground. One perched in a bush and preened, first under one wing, then the other, all the while flashing his dark orange crown patch.

The golden-crowned kinglet is almost as small as a hummingbird and can even hover like one. This tiny olive and buff-gray bird has white wing bars, a whitish eyebrow stripe, and yellow feet. Named for the golden crown of the female, its alternate names, such as "flame-crest" and "fiery-colored wren," refer to the male's dark orange crown patch. Both male and female crowns are outlined in black. Their scientific name *Regulus satrapes* is a combination of Greek and Latin, meaning "small king" and a "ruler who wears a golden crown."

Studies indicate that golden-crowned kinglets winter throughout southern Canada and south across the United States into Mexico and Guatemala, in both coniferous and deciduous forests. They are affected by severe cold and snow, so their winter densities are highest where temperatures never fall below freezing. According to James Ingold and Robert Galati's account of golden-crowned kinglets in *Birds of North America*, they seem to lack the lipid reserves needed to maintain body temperature during cold winter nights, and may conserve energy by taking cover in vacant squirrel nests. We certainly have plenty of those.

The winter migrants usually appear in Pennsylvania in early October, and they often forage in mixed-species flocks during the fall and winter. Throughout the early autumn, I have seen them associating with yellow-rumped, palm and black-throated blue warblers and blue-headed vireos. But their late fall and winter associates are predominately brown creepers, along with downy, hairy, and red-bellied woodpeckers, black-capped chickadees, white-breasted and red-breasted nuthatches, tufted titmice, and dark-eyed juncos.

Some time in late March to mid-April most golden-crowned kinglets head north to breed in the boreal spruce-fir forests. They particularly favor spruce and tamarack bogs or mixed stands of balsam fir, white spruce, and red pine, according to Robert and Carlyn Galati, who first studied their nesting behavior in Minnesota's Itasca State Park from 1954 to 1957 and in 1960. The Galatis spent 3,800 hours observing nineteen kinglet nests involving thirteen different breeding pairs.

Luckily the Galatis did not suffer from vertigo, because they discovered that on average, kinglets nest fifty feet off the ground. To study the species' nest life, they spent seven days a week from dawn to dusk sitting in tall wooden towers built next to nest trees.

Before this remarkable treetop study, nothing was known about territorial behavior, incubation period, or nestling and fledgling stages of golden-crowned kinglets. The Galatis discovered that their kinglets' nests were built in upper tree crowns close to the trunks, well-protected from the wind and rain by overhanging foliage. Both parents construct the deep, cup-shaped nests out of mosses, spider webs, and cotton from cotton grass, and then line them with lichens, birch bark, deer hair, and feathers. During the nest-building period of four to six days, the pair also engages in courtship feeding.

Only the female incubates the five to eleven, white-to-cream-colored eggs speckled with pale brown and faint lilac, while the male sings and accompanies her whenever she leaves the nest. He also defends his territory, an average of four acres, from other kinglets, and harasses potential predators such as red squirrels, blue jays, and gray jays. After fifteen days, the eggs hatch and both parents feed the young. Halfway through the nestling period of sixteen to nineteen days, the female starts building a new nest and lays a second set of nine eggs. Meanwhile the male continues feeding the first nestlings, which remain dependent for another two weeks. The second brood gets the full attention of both parents.

Golden-crowned kinglets were once considered a rare breeding bird in Pennsylvania. Historically, they nested only on the Pocono Plateau and in a scattering of north-central counties where there were large red or black spruce trees. After those areas were clear cut at the turn of the century, the only known site for nesting golden-crowned kinglets was a remnant of uncut spruce forest near Pocono Lake.

But in 1972, nesting golden-crowned kinglets were discovered at Wild Creek Reservoir in Carbon County, slightly south of their historical range in Pennsylvania. They were using a nonnative coniferous plantation, just like red-breasted nuthatches, that was established in the early 1900s on cutover lands newly purchased by the commonwealth's Department of Forests and Waters.

By the time that survey work for the first Pennsylvania Breeding Bird Atlas Project finished at the end of the 1980s, nesting golden-crowned kinglets had been located in many mixed Norway spruce and red pine plantings far from their historical range. Many of the birds preferred the higher elevations found in the Appalachian Plateau and Ridge-and-Valley provinces, but planted spruce and pine farther south in the Piedmont Province also had confirmed nestings. A few nests were even found in small, coniferous groves in wooded residential areas.

Twenty-seven years after we planted our Norway spruce grove at the top of First Field, I was astounded to hear the thin, high song of the golden-crowned kinglet coming from it in May. After a little searching, I found not only a pair of kinglets, but the nest they were building. It was not fifty feet high in our spruce trees because they were not that tall. Instead, it hung near the end of a branch twelve feet from the ground beside a trail, well-hidden, I thought, by the drooping, well-needled branch. I watched them from beneath a nearby spruce tree as they finished building their nest and then during the egg-laying and brooding period. He nearly always accompanied her when she left to forage, and sang both when she left the nest and when she returned. I didn't go near the nest for fear of attracting predators, so I didn't know how many eggs she had, but I did see the pair bringing in food at the end of June.

Since this was the first known nesting of golden-crowned kinglets in our county, I was excited. But a couple of days later, there was no sign of activity around the nest. Instead, I heard the male singing at the other end of the grove. Finally, I checked the nest. It was empty, and we thought that the American crows, which had nested in the grove the year before and had just fledged young in nearby woods, were probably the culprits.

I followed the singing of the male and he practically led me to their next nest. It was a mere ten feet from the ground, underneath another drooping spruce branch, and didn't seem to be as well built or well hidden as the first one. Once again I settled down under a nearby spruce tree to watch. Once again, just as they hatched, at the end of July, the nest was empty. This time the adults disappeared too. Whether the nest predator had gotten them also or they had given up and left for a safer place I never knew.

DECEMBER 29. More snow showers today. The sharpie was back by 8:30 A.M. in the lilac bush beside the front porch. Dave and I stood outside watching it, fluffed up and forlorn, while the black-capped chickadees, on full alert, warned off all the other birds. At 9:15, when I went outside, the feeders remained empty and the birds were still, so I was certain the sharpie was hiding somewhere nearby. Poking about in the yard, I flushed it out of the juniper tree and it flew to the edge of the woods where it sat defiantly waiting even as I set out on my walk.

Along Laurel Ridge Trail, I heard a rustle in the snow-covered, fallen leaves and two tiny masked shrews emerged to chase and squeak for a couple seconds before dashing under cover. That was the first time I've ever observed shrew-chasing in midwinter. Usually, I see them chasing and squeaking in early spring or midsummer and had assumed it had to do with mating. But perhaps it is also the way these solitary creatures defend their turf.

Snow continued sifting down and I sat beneath the sheltering spruces. Red-bellied, pileated, and hairy woodpeckers were the only birds I heard.

DECEMBER 30. I basked against the Far Field Road embankment between two fallen logs, soaking up the sunlight this bright, warm day. Then I heard rustling noises in the thawing leaves nearby. At first I dismissed them as foraging gray squirrels, but it didn't sound quite right. Slowly, I eased into a sitting position and tried to find the source of the noise. Two hen turkeys were scratching vigorously in exposed areas of leaf duff about a hundred yards away. A third turkey crossed the road, having come up from Roseberry Hollow. After looking in my direction, she joined the pair.

I thought she had seen me and would warn the others. Instead all three continued raking the ground with their feet and bobbing their heads down quickly to eat whatever they found, moving slowly up the ravine and back and forth. Gradually, they made their way to the edge of First Field and finally ambled off. They never did see me.

Later, walking up the road, I noticed the changing complexion of the stream. In several areas the water flowed below a thin, translucent skim of ice, looking like shoals of fish or giant amoebas undulating below the surface. Kaleidoscopelike, the shapes continually shifted as I watched.

DECEMBER 31. On this last day of the millennium, I walked down the hollow road and imagined what it would have been like if I had been transported back to the early nineteenth century, before Europeans had arrived to cut the trees for charcoal. The road would not have existed, the trees would have been bigger, and the species different: for instance, the American chestnut was still a mighty tree, producing food for an abundance of wild creatures. Cougars and wolves roamed the mountain. The forest was colder because hemlocks were the dominant trees along the stream. Our mountain was then unbroken forest visited only occasionally by native Americans.

This Sunday morning, within the depths of the hollow, no sounds from our technological society reached me so it was easy to pretend that I had entered a time warp. All I heard above the flowing stream were American goldfinches, a red-tailed hawk, black-capped chickadees, and golden-crowned kinglets. But sitting on a modern bench brought me back to reality.

Putting benches beside our trails, as my husband Bruce decided to do, struck me as an admission that we were getting older, an uncomfortable acknowledgment that our time on earth was running out. I preferred to sit at the base of a tree, seemingly at one with nature.

"I *am* getting older," Bruce insisted, "and I want benches to sit on."

So, late in the winter of '99, Bruce consulted with two hunter friends, Jeff Scott and Bob Myers, who are also talented carpenters. The deal was that we would purchase the pressure-treated lumber and Jeff and Bob would build the benches, assisted by Bruce and Dave. As it turned out, Bruce and Dave mostly kibitzed while Bob and Jeff turned out—in record time—four handsome benches, complete with slanted backs and armrests.

By the first week in April, after much discussion of where we should put them, Bruce had installed three of the four benches. At first they were named according to their locations—Far Field Road Bench, Ten Springs Trail Bench, and Hollow Road Bench—but within a week they had new names.

The first to be renamed was the Ten Springs Trail Bench, which Bruce had installed at the edge of the clear-cut. It overlooks the mature uncut deciduous forest of Laurel Ridge.

On a warm, sunny April morning, I walked through Margaret's Woods to the clear-cut along Greenbrier Trail, following the sound of a gobbling tom turkey. Using my old Lynch Foolproof box call, I tried to entice him into view. Finally, I glimpsed him ahead of me on the trail, slowly fanning his tail feathers. Even though I stood absolutely still, he spotted me after several minutes, stared as if in disbelief, and then ran off up the slope.

From Greenbrier Trail, I hiked down Dogwood Knoll to Ten Springs Trail and the new bench. As I sat there, I again heard a turkey gobbling and answered him with what I thought were enticing hen calls. We called back and forth as he moved unseen up Laurel Ridge. I never did see him, but that was how Ten Springs Trail Bench became Turkey Bench.

Two days later, on another sparkling spring morning, I walked down the road to Hollow Road Bench. I was just in time to watch and listen to two male Louisiana waterthrushes singing over prime Louisiana waterthrush habitat. These wood warblers are some of the earliest returning migrants and, with their brown backs, white breasts streaked with brown, and long pink legs, they look more like thrushes than warblers. Because they favor stream banks in wooded ravines for nesting, our hollow usually hosts three or four pairs.

At one point they stood on two fallen trees, a couple of feet from where I was sitting, and chipped, swayed, and wagged their tails up and down as Louisiana waterthrushes are so often wont to do.

I remained still, and they continued their performance, flying up and down above the stream, singing their melodious, ringing song, emitting their sharp, metallic "chip" calls, and chasing one another, though I saw no real fighting. One waterthrush waded through the rushing water, feeding and singing, reminding me of a dipper I had once watched in a Wyoming stream. Needless to say, Hollow Road Bench was renamed Waterthrush Bench and continued to give me a front seat view of Louisiana waterthrush behavior throughout the month.

It took another four days, until April 14, before the Far Field Road Bench received its new name. The previous evening the phone had rung at 7:30. My elderly father's even older neighbor, John Stine, had some bad news. My father had fallen outside his country home while working in his garden and had lain for hours on the cold April ground, unable to get up, and calling for help. Luckily, Mr. Stine had glanced out of his window and noticed my father lying there, several hundred feet away. We had spent much of that night at the hospital, learning, finally, that Dad had broken his hip.

The following morning I had a difficult time taking my usual morning walk because I was worried about Dad. I was wondering why he had broken his hip less than a year after he had fallen and broken his leg. But I forced myself outside that windy, cold day and headed for the Far Field Road Bench. Set along my favorite refuge on the property, it had already become the bench I used the most. I sat and listened to a singing blue-headed vireo. Looking downslope into Roseberry Hollow, my mind was so busy with questions that I paid little attention to the scene around me.

What if Dad's neighbor had not gone to the window? How much longer could Dad live alone in his beloved country home? Would he survive his hip operation?

A slight movement to my right aroused me from my reverie. The coyote sensed me at the same moment that I sensed him. He was less than ten feet from the bench and had evidently been moseying along the old woods road as inattentive to his surroundings as I had been to mine.

He was a magnificent, full-grown male who looked me fully in the face for several seconds before turning around and loping slowly away. I don't have a mystic bone in my body. Yet when the coyote appeared so unexpectedly, I felt as if an unseen hand had blessed me. I remembered that many native Americans had venerated the coyote. The Crows called him First Worker, creator of the earth and all living creatures, and desert southwest tribes referred to him as God's dog. Others called him "Gray Brother" and "Shining-Eyed Companion."

Did the coyote know I needed a revelation? Or was his appearance, the first ever I had had of an adult coyote on our mountain, merely a coincidence? Surely the latter, my rational, scientific mind told me. But Dad did come through his hip operation beautifully, and the prognosis for a complete recovery was excellent. Immediately, I christened the Far Field Road Bench Coyote Bench in honor of God's dog.

The fourth bench remained in the barn until mid-June. A

twinge in my back reminded me that sometimes, when my back gives me trouble, I can't walk as far as I usually do. So I suggested that we put the bench in the Magic Place, a short distance from our house. Nestled amid the old-growth red and black oaks, it's the only bench that is not on a trail.

For less than a month we called it the Magic Place Bench. But one day in July, my friend Colette Heller and I sat there and watched a female box turtle watching us. Several weeks later Bruce was eyed by the same box turtle. "Let's call it Turtle Bench," he suggested. And we did.

Our next bench was a memorial bench built to honor one of our youngest hunters. Seventeen-year-old Alan Harshberger died in a pickup truck collision, through no fault of his, on Memorial Day weekend in 2000.

The bench was built by Tim Tyler, with Jeff's instruction. Tim, a hunter friend of ours who is a close friend of the Harshbergers, situated the bench at the top of First Field tucked back in our Norway spruce grove. There it overlooks the knoll where Alan had his hunting stand. From Alan's Bench we can see fifty-five miles on a clear day and we name the mountains—Nittany Mountain, Egg Hill, Tussey Mountain, and our own Bald Eagle Ridge heading northeast to Williamsport.

Recently, Jeff found some "scraps" and built us still another bench. Again, thinking of my aging, I had it placed halfway along Lady Slipper Trail. To reach it I walk half the Black Gum Trail, then climb up to the bench, and on to Laurel Ridge Trail. So far, I have not renamed it although I suppose I could call it Fox Squirrel Bench since I sat on it early in December and watched a fox squirrel climb up a nearby tree and look me over.

This is a bench that is easily reached on snowshoes and I often snowshoe along the trail when the conditions are right, as they have been most of this unusually cold and snowy December.

January

January observation can be almost as simple and peaceful as snow, and almost as continuous as cold. There is time not only to see who has done what, but to speculate why.

Aldo Leopold

JANUARY 1. A frosty earth was lit by the rising sun that poured down as a benediction to the last century and a welcome to this one. As the century wound down last evening, I had the distinct impression of emerging from a tunnel into a new life, leaving behind the good, who died too soon, and the evil, who didn't—both the famous and infamous as well as those unsung folks who merely lived and did their duty but did not make it into a new century—Bruce's Dad and Mom, my Mom, my Nanna and Poppop, my Great-Aunt Mary, my Aunt Marge, and all of Bruce's aunts and uncles except Uncle Gilbert. Of our parents, only my Dad has outlived the century he was born in, but all of our siblings and mates, except for one sister-in-law, our children, and theirs will have lived in two centuries.

Listening to commentators praise and curse the twentieth century, its triumphs and excesses, the people who made a difference for good and evil, the Hitlers and Einsteins, the Stalins and Gandhis, I was struck by the fact that it was all human-centered. Not a mention was made of the Earth that continues to sustain us or the people who started the environmental movement. We are so caught up in the technological revolution, the vast entertainment industry, the wars, and intervals of peace that I fear we are entering this century with even more hubris than we had at the beginning of the last century. When will we seriously address the issues of global warming, air pollution, dwindling clean water supplies, dwindling fish stocks in the ocean, the dwindling of wildlife everywhere, and the mass extinctions that threaten a catastrophic loss of global biodiversity?

As I set out on my walk, accompanied by the calls of wintering robins, watching and hearing jet planes overhead and cars and trucks in the valley, I wondered what my descendants will be hearing and seeing at the beginning of the next century. Surely, like my own ancestors, I cannot begin to imagine the changes that will occur over what is, in reality, a blink of time in the life of a universe too vast to comprehend. But I can hope that we will attain some measure of peace with each other and with the Earth itself.

I stood quietly, contemplating the Far Field Thicket as the sun flashed in and out and the snow sparkled in the intermittent light. First one, then another, ruffed grouse erupted from cover. Altogether five emerged from the grapevines at the base of several trees. One was close to where I stood, yet I never saw it until it broke cover.

The Far Field Trail was a deer highway. So was the field itself. It was hard to believe that the hunters have taken more than thirty deer off the property.

By the time I reached the spruce grove, a light snow was falling. As I sat on Alan's Bench, I was spangled with six-pointed stars, a blessing for the new millennium.

I poked about in the grove and found a scattering of rosy-pink and gray female northern cardinal feathers. Even though

I discovered no hawks or owls, something had clearly had itself a meal there not too long ago.

JANUARY 2.   I listened to the glorious prologue to Boris Godunov on the Metropolitan Opera broadcast and then, unwilling to listen to the male angst and evil that constitutes the rest of the opera, went out into the natural world, where evil does not exist. It is the only antidote to the continual dismal human news of slaughter and hatred that clog the news reports.

From the top of the mountain, I looked across at a patch of sunlight illuminating the village of Birmingham like some nineteenth-century painting, tucked among the forested hills, the foreground a field of Amish haystacks lined up four-by-four with geometrical precision, a weathered, faded barn in the background. From there I started down the hollow road. A large flock of American crows called vociferously as they congregated on tree limbs in sunny places above the road, keeping one sunny spot ahead of me. Near Waterthrush Bench black-capped chickadees and American goldfinches harvested hemlock seeds that pattered down like hail. Some golden-crowned kinglets also foraged for insects in the hemlocks and the deciduous trees. On the ground a flock of dark-eyed juncos ate black birch seeds.

I paused to watch a deer cross the road in a patch of sunlight and move cautiously up the ice-covered slope. Snow beams drifted past in the sunshine. The snow covers a multitude of ugliness elsewhere, such as the trash dumped at the bottom of the mountain, but in the hollow it merely graces what is already beautiful.

JANUARY 3.   An eastern bluebird called when I stepped outside in the early morning warmth. During my walk, the half-frozen earth oozed beneath my feet. I sat under the one large white oak left on Dogwood Knoll and listened to the rustle of its still-brown, dried leaves clinging to the branches. A band of white-gold over Laurel Ridge broke the monotony of grayish-white skies.

I've been reading Kathleen Norris's *The Cloister Walk,* which is about the attraction of monasticism and community, where life is simpler and monks must live harmoniously in isolated settings, worshipping and praising God. Separated from the temptations of our complicated world, they are allowed many hours to read, help others, sing, and worship. Most Americans are baffled by their choice.

Similarly, our community of three on this mountain is an alien form of living to most people. Good food is important to us, sharing meals, talk, and ideas, reading, writing, and spending hours outside in work and contemplation. Our only concessions to the modern entertainment/advertising world are using our computers, listening to our public radio station, and our small CD collection of mostly twentieth-century classical music, and reading nature, conservation, and business magazines, and one international newspaper. During the winter we are especially isolated and make only occasional forays out to pick up mail, buy groceries and birdseed, visit my father, do research in the university library, attend Audubon meetings, or take in an infrequent concert or movie. But mostly we are here together, living harmoniously in the deep silence of winter.

Today, though, was Dad's eighty-eighth birthday and we planned a surprise party for him. Sixteen people attended—all four of his children, five of his ten grandchildren, his two great-grandchildren, his two sons-in-law, one of his daughters-in-law, one of his granddaughters-in-law, and one of his grandsons-in-law. Not a bad turnout for a weekday. Eva and her cousin Patrick were especially excited as we hid in the dining room of his assisted-living facility while Dad made his slow way toward us using his walker and accompanied by the social director. She had told him he was being transferred to another dining room for the day to make room for two birthday parties for two other men in the facility whose birthdays are also on January 3.

We had decked the room with a banner and balloons, the children kept peeking from behind a plant to report his progress toward the room, and finally, when he reached it, we all yelled "Happy Birthday!"

Eva hung over him and supervised his opening of mostly gag gifts. She was also fascinated by the cake, which was covered with facsimiles of photos from all stages of his life—childhood, college, adulthood, and old age.

Afterwards, we went over to his country home to resume festivities for the afternoon and evening. Dad especially enjoyed watching the children and playing bat-the-balloon with adults and children. While he had to remain seated because he had fallen and broken his second hip last fall, he could use his arms and seemed to enjoy it.

"Were you surprised?" I asked him.

"I was astounded," he said. "It's one of the happiest birthdays in many years."

Reflecting on it, I realized that he had probably never had a birthday party, let alone a surprise one. Another five-star day for the holiday season!

JANUARY 4. A clear, cold morning again and I walked down the hollow road. At the big pull-off, the woods were filled with the usual songbirds—brown creeper, winter wren, golden-crowned kinglets, dark-eyed juncos, white-breasted nuthatches, black-capped chickadees, and tufted titmice. One singer up the slope, though, sounded different and I could not pish it down to verify, but it sang twice. If it were not the dead of winter, I would think it was a blue-headed vireo with its ringing song and pauses in between. Is that possible?

The stream was frozen solid in most places and the rhododendron leaves rolled as tightly shut as cigars in the fourteen-degree cold. Occasional flakes sifted down through the hemlock trees.

I heard two thumps on the bow window shortly after noon and glancing out at the feeders saw not one bird.

"There must be a predator out there," I told Luz. Sure enough. A sharpie sat on a sapling in the flat area, twitching its tail. We had a great view of it before it flew off empty-taloned. Except for that incursion, the feeders were mobbed all day. At one point I counted at least thirteen American goldfinches and dozens of dark-eyed juncos. Eva spent a good deal of time looking out the bow window at them and was able to identify black-capped chickadees and northern cardinals.

I tell her the northern cardinals are Christmas birds. Resplendent against the snow, the male glows like a Christmas light. What is he doing here among his drably suited brethren? Why isn't he in the tropics at this time of year with the other gaudily attired birds?

At one time the northern cardinal was a bird of the South. John James Audubon knew it as the "cardinal grosbeak" of Louisiana and South Carolina, and declared "in richness of plumage, elegance of motion, and strength of song, this species surpasses all its kindred." Most of us in the North would agree with his assessment, especially at this time of year.

James Lane Allen immortalized the bird in his book as the "Kentucky cardinal," and Virginia claimed it as its own with alternate names such as "Virginia redbird" and "Virginia nightingale." Henry David Thoreau never saw it around his Concord, Massachusetts, home, because it wasn't until 1958, nearly a century after his death, that northern cardinals began nesting in Massachusetts.

Here in Pennsylvania, northern cardinals were common in the southeastern and southwestern corners of the state by the late 1800s. Gradually, they moved north along the river valleys, reaching central Pennsylvania by 1912, Crawford County in 1928, and occupying the rest of the state by 1960, although they are scarce in heavily forested north-central Pennsylvania.

Today, northern cardinals live throughout eastern and central North America, from southern Canada to northern Guatemala. Seven states claim the northern cardinal as their state bird—Indiana, Illinois, Kentucky, North Carolina, Ohio, Virginia, and West Virginia.

The northern cardinal is the quintessential generalist, living successfully in a wide range of habitats heavily impacted by humans. Instead of retreating when eastern forests were cut, northern cardinals advanced northward, living in dense shrubbery planted in hedgerows and yards, and actually preferring to forage on town and suburban lawns. They also reside in shrubby, logged, and second-growth forests, shrubby grasslands and marsh edges.

Once the eastern forest cover was removed, the climate became warmer, which also encouraged cardinals to travel north. Then people started feeding birds in the winter, a final boon to north-moving northern cardinals, who can survive an average minimum January temperature of five degrees Fahrenheit.

Anyone who feeds birds can attest that northern cardinals come early in the morning and late in the afternoon on most winter days. The males are dominant, frequently chasing females from food in late fall and early winter. Many northern cardinals remain mated for life, but they often join flocks of juveniles and adults in early autumn. Membership fluctuates, averaging between five and twenty birds, and consisting of an equal number of males and females. As it gets colder, the size of the flock increases, especially if there is abundant food and cover. Sometimes, northern cardinal flocks loosely associate with other species, such as dark-eyed juncos, white-throated sparrows, tufted titmice, song sparrows, American tree sparrows, and American goldfinches.

Here on our mountain, favorite winter northern cardinal feeding areas are thickets of greenbrier and wild grapevines on sheltered south-facing slopes. No sight is lovelier on a winter day than that of a flock of northern cardinals against an azure winter sky eating grapes from vines high in the tree canopy.

On cold winter nights, northern cardinal body temperatures drop three to six degrees, and they roost together in thick shrubbery or conifers to conserve heat. Our spruce grove is a favorite spot. Their major avian predators are Cooper's and sharp-shinned hawks and eastern screech-owls. The latter sometimes kill them on their nighttime roosts and may be the culprit in the killing of the spruce grove northern cardinal. Other major predators are foxes, feral cats, minks, and weasels.

Northern cardinals begin singing in the thickets by the third week in February. They engage in mate-feeding, an activity in which the male picks up a seed, hops over to the female, and, as she takes the food, they briefly touch beaks. Both sexes sing and engage in bouts of "countersinging," when first one bird, usually the male, sings one phrase several times and then the other matches it. This type of singing is thought to synchronize and unify cardinal couples. When it's practiced between males, it helps to settle territorial disputes over each other's two- to ten-acre territory.

Northern cardinal songs wax and wane throughout the breeding period into August. Before new pairings take place, males may sing 150 to 200 or more songs in the dawn light. Females sing mostly before nesting or with the male through nest building. Males continue singing to a lesser degree during incubation by the female, and sing even less when feeding nestlings and fledglings.

The male, after all, is a busy, involved "husband" and "father," feeding his mate every four or five minutes before nesting and once a minute when she is busy building her nest and laying her brown-speckled, olive-white, three to four eggs. He continues feeding her during the twelve- to thirteen-day incubation period, and is the major provider of food for the nestlings and fledglings.

The female constructs the four-layered nest of stiff weed stems, leaves, and/or plastic, grapevine bark, and rootlets in thick shrubbery, four to seven feet above ground at the edges of woods, in hedgerows or fencerows. Her favorite nesting shrubs are all alien species—Japanese honeysuckle, multiflora rose, and privet—although she will also use dense evergreens and native shrubbery.

The male accompanies her as she builds the nest, probably to protect her from predators and other males that might be interested in some hanky-panky. The nest building takes as little as three and as long as nine days to finish.

Most northern cardinal nesting attempts are unsuccessful because snakes, small mammals (particularly chipmunks and squirrels), and birds such as blue jays prey heavily on both their eggs and nestlings. But once they fledge, anywhere from 60 to 80 percent survive to adulthood. Fortunately, they have a long breeding period, from April 3 until August 16 in Pennsylvania, so they can produce as many as eight clutches. The northern cardinals' motto should probably be, "If at first you don't succeed, try, try again."

When the eggs do hatch, the young remain in the nest ten to thirteen days. After fledging, they are dependent on their parents another forty days. Mostly, they are fed insects, even though the annual diet of northern cardinals consists of only 29 percent insects and 71 percent fruits and seeds. Northern cardinals eat at least eighty-five different insects and seventy-seven plants. They especially like blackberry, raspberry, and dogwood fruits, and the seeds of wild grapes, smartweed, and bindweed. Other seed sources include sedges, foxtail, vetches, dock, sumac, vervain, and tulip trees, as well as corn, oats, and oil sunflower. They also like the buds of trees, particularly elm and chokecherry, and they even drink sap from yellow-bellied sapsuckers' tree wells.

Once the young learn to forage for themselves, they disperse, probably no more than a mile from their parents' territory, although a few banded young have been found as far as one hundred miles away. By December they look just like their parents, and as soon as the adults start singing again, the juveniles listen, imitate, and finally learn to sing as beautifully as the old adults by April.

To me, northern cardinals epitomize a dignified beauty. That's why the great Swedish taxonomist Carl Linnaeus named the birds for red-hatted and robed cardinals of the Roman Catholic Church—*Loxia cardinalis*. After several more changes, today its scientific name is *Cardinalis cardinalis*, as if to emphasize its aristocratic bearing.

JANUARY 5. Lightly snowing at dawn, and so I first walked down the hollow road to see the hemlocks draped in white. Just above the hemlock-dominated forest, dozens of golden-crowned kinglets and black-capped chickadees foraged close to the ground and a winter wren called and searched for food beside the stream.

Then, on the last slope, near the bottom of the mountain, I listened to a pair of Carolina wrens duetting. One was high up on the slope, the other near the stream, and each echoed the other.

Coming back up the road, I heard again what could only be a blue-headed vireo singing. He sang twice with a long gap in between so I had to sit and wait for that second song. But I couldn't spot the singer for a positive identification. Blue-headed vireos do eat some fruit and many winter in the southern states so it is possible that one decided to stay and live on the bonanza of wild fruits here. In this weather, though, I doubt it will survive long. Unfortunately, the slope is too steep and icy for me to climb up and look for it.

Later, along the Far Field Road, I heard the loud calling of a northern cardinal in distress and directly in front of me one twisted away from an attacking sharp-shinned hawk by evasive, fast, fluttering flight, screaming all the while. Consequently, the sharpie went without a meal and the cardinal went free.

JANUARY 6. A warm thirty degrees and mostly clear. I was outside by midmorning on a walking meditation. The Far Field Road and slopes below rustled with scratching American robins. They called, along with cedar waxwings. Then I heard a song that I wrestled to extract from my memory. It sounded like an upside-down eastern meadowlark song, an impossibility at this time of year. And then I remembered. A singing brown creeper was somewhere below me. Never before have I heard a brown creeper singing in the winter.

I sat on Coyote Bench, as the clouds gathered and the sun dimmed, watching more and more robins feeding both on the melted ground and in the grapevine-draped trees. Robins landed on the ground and on nearby branches, surrounding me on the bench, and watching me as I watched them. I dared not move as wild grapes plopped down around me. Next a few European starlings joined the party, along with a northern flicker.

Sitting there in the warming sun, encircled by the birds of spring in a winter woods, the ground still patched in white, seemed like a winter mirage. But it was real enough. Finally, though, after half an hour, most of the American robins and European starlings moved on up over the mountain and I was left with the calls of the usual winter birds—black-capped chickadees, tufted titmice, white-breasted nuthatches, and a red-bellied woodpecker.

Walking back, I saw dozens more scratching American robins ahead on the Far Field Road and below me. What a sheltered cornucopia Roseberry Hollow is! Then I spotted one of at least two calling northern flickers. He was poking into a woodpecker hole high up on a tree trunk. The other flicker foraged on or near the ground, flicking the dried leaves aside with its long, pointed bill.

As I rounded the last curve on the road, more robins and

flickers appeared, but during most of the rest of my walk, the woods and First Field were nearly empty of birds, except when I walked on First Field Butterfly Loop, below the warmed, open, Sapsucker Woods slope. There I heard and saw more foraging birds—American robins, dark-eyed juncos, a red-bellied woodpecker, another brown creeper, and pileated and downy woodpeckers.

Not a bad bird count for early January nor a bad day for a walking meditation. What would I do without birds to lighten every winter day with their exuberance?

JANUARY 7. I learned a lesson today. Never stay in, no matter how unappealing the weather might appear from the inside looking out. Heavy fog and a light mist had crusted the patchy snow cover overnight and visibility was limited to a couple hundred feet. Shades of gray seemed to be the only colors left in a world gone dully monochromatic.

But Bruce and I craved fresh air and exercise, and so we forced ourselves to go walking at midday. About three-quarters of the way along the Far Field Road, the sodden scene was suddenly transfigured. Weather conditions had conspired to rim the bottom edge of every tree branch and vine with white, needle-width, inch-long deposits of rime frost, all of which curved toward the south or down the mountain slope. Evidently cold air seeping from the north off Sapsucker Ridge had met the thicker fog roiling up from Roseberry Hollow and turned the woods between the First and Far Fields into a frosted wonder. We kept craning our necks to peer into the tallest treetops, which resembled enormous, old-fashioned, white, feather dusters.

Then, as we returned along the Sapsucker Ridge Trail, where we discovered that the frost points were even longer— $1\frac{1}{2}$ to 2 inches—I glimpsed, over in a protected spot, the flash of a crow's wings as it flew up from the trees. But when I scanned the area, instead of more crows I spotted a large, plump object sitting out on a tree limb a hundred feet from where I was standing.

I fiddled with my binoculars, trying to see clearly in the thick fog, and called softly to Bruce, who was several yards ahead of me, to move quietly and come see what looked like an owl. Expecting it to be a great horned owl, I was surprised that it stayed seated while we gawked at it, since great horned owls are usually off in an instant when discovered.

But this was a rarer species, one that we have glimpsed only occasionally during our years here—a barred owl and no doubt a female because it looked immense. She sat, a puffy vision decked out in white and gray feathers, while tendrils of fog curled about her rime-frosted tree perch in that silent, ghostly woods.

Breathlessly we watched as she turned her head back and forth on her swivel neck, then bent to peer down at the ground beneath her. Either she was unaware of us or, more likely, unconcerned, since barred owls are known to be the boldest of owl species. They also embody the essence of owlishness, with their slow, deliberate movements, their neatly rounded-off, soft-looking heads framed with what looks like enormous fur muffs, and their dark, liquid eyes accentuated by large, oval, facial discs. The horizontal barring around their collared necks and the vertical barring of their breasts are other distinctive identification tags. Altogether, barred owls are thoroughly handsome, satisfying owl creatures to watch in a shrouded wood.

I knelt, one knee in the icy snow, and studied her for at least ten minutes before she turned her back and I could see some black mixed with brown on her stubby tail. Then she leaned out and slightly down before launching into silent flight that propelled her out of our sight.

Immediately an unseen pileated woodpecker began hollering ahead of us in the woods. We kept walking and peering into the misty treetops looking for a sign of the barred owl, thinking that it had, perhaps, grabbed the pileated's mate. But barred owls can handle no bird bigger than a flicker. Pileated woodpeckers are known to make a crowlike ruckus and sometimes dive at barred owls when they are in the vicinity. Whatever may have been the case, we did not see the owl again, but after several screaming minutes the pileated swooped low over our heads, still yelling.

We left that frosted wood then and were immediately propelled back into a gray, sodden world, grateful to have had, for a while, a taste of winter enchantment and the best view ever of a barred owl.

JANUARY 8.   Six degrees at dawn and partially clear with another skim of snow on top of the icy snow. Again I was drawn down the hollow road because it is a refuge for so many birds—a dozen or more black-capped chickadees in the hemlocks along with a brown creeper, golden-crowned kinglets, tufted titmice, downy woodpeckers, a northern cardinal, dark-eyed juncos, and white-throated sparrows. One junco was so intent on scratching up a patch of exposed dirt that it came within a couple feet of me. Best of all was the sight of a hermit thrush at the first pull-off near the frozen stream. It, like the blue-headed vireo, should have gone south but had been tempted to stay by the wild fruit crop. I heard no more blue-headed vireo songs, though, so either it is gone or dead from the cold.

In the evening, Mark spotted a gray phase eastern screech-owl sitting on the roof of the wooden bird feeder while one frightened dark-eyed junco flew back and forth under the porch roof. The screech-owl seemed supremely disinterested in the bird, probably because its real intent was to catch mice or voles attracted to spilled feeder seed on the ground below. The owl gave all of us, including Eva, a long, satisfying look before finally flying off, and Eva spent the rest of the evening talking about the little owl and reenacting its flight.

JANUARY 9. After a heavy snowfall overnight, this morning the sky was more blue than gray, the sun was shining, and a breeze had sprung up. I sat on Coyote Bench listening to American robins calling in Roseberry Hollow and watching clumps of snow cascading from tree branches in the warm sun.

Mark asked us, the other night, how we felt about death and the possibility of life after death and Bruce said that he never thought about it. I do, though, as I watch my father slowly decline physically and mentally, yet cling to his diminished life stubbornly, finding happiness in a living facility he would have scorned years ago even though it is the best of such worlds, its lavish decor and grounds like a well-groomed country club, its staff friendly and caring. Although he is a practicing Christian, faithfully attending the Sunday afternoon services that his living facility provides, when Mom died, he sighed and said he hoped she was right about life after death, but it was clear that he had his doubts.

I think of death only in terms of how much longer I have to enjoy the wonderful life I have here on this beautiful mountain with my best friend and lover, Bruce, and our son Dave. I give thanks for every day I can be outside and moving around in a semblance of how I moved when I was in my thirties. I want to be a fit old woman and sharp enough mentally to keep writing and reading challenging material. My inspiration is my college botany professor—Dr. Wayne E. Manning—now 103, still walking half a mile every day along the Susquehanna River in Lewisburg, still able to travel with friends, still lighting a candle in church on Christmas day, still living in his own home with the help of friends and neighbors. He's almost forty-two years older than I am. At that rate I still have nearly half a lifetime to live.

I followed an avenue of deep, porcupine tracks from the edge of Sapsucker Ridge to the upper edge of the spruce grove. There was only a little fresh gnawing on one tree trunk; the group of three trees there was already strangely shaped from former porcupine prunings in winters past.

It was forty-four degrees by midafternoon. The snow was melting fast. Near dusk an opossum followed Dave's path up across our lawn from the guesthouse, its nose close to the ground. But every time it reached the last steep curve around a black walnut tree, it turned around. I watched it advance and retreat three times until it went down to the road ditch where Dave encountered it. He thought it was sick because it made no menacing moves or sounds and didn't play dead. Later it was gone, so who knows. Now that we are in deep winter, the triplets and opossum have not been coming to the feeders at night.

JANUARY 10. At 8:50 A.M., I noticed the female downy woodpecker pressed against the inside of the bird feeder, her bill pointed upward in frozen fear. The sharpie sat nearby, but it soon flew off and the birds resumed their feeding.

I walked in slushy snow through Margaret's Woods and paused, as I often do, to look at the fading remains of her home. She was related to the family who owned our home for almost a hundred years and lived with her invalid brother in her parents' home until he died. Then she continued on alone and our boys often visited her, especially during holidays. She entertained them with stories of the mountain when she was a youngster.

Years before she had lost her home in a sheriff's sale and had gotten a friend to buy it for her so she and her brother could continue to live there. It was the friend who, needing money to send his son through college, sold Margaret's home and one hundred acres to a logger. In the midst of the logging, she died, and in her honor I named the woods the logger kept around her house "Margaret's Woods."

Off Greenbrier Trail, a porcupine fed high in a medium-sized chestnut oak tree. It was actively biting off small branches as it moved higher into the canopy. It had its back to

me, but judging from its slight movement, it was also gnawing on the bark as the branch swayed in the breeze. Robins
called below while I watched the porkie.

Finally, I moved directly below its tree and noticed that its
tracks had come from farther up the ridge. A scattering of
twigs on the ground beneath the tree had been gnawed on
briefly before being discarded by the creature. Slowly it
inched itself higher on what looked like a dangerously thin
branch. One front claw was wrapped around the branch as
the porkie chewed on it and slivers of bark fell to the ground.
Of the many winters I have spent watching porcupines, this
was one of the best views I have ever had because it was un-
knowing and/or uncaring that I sat below.

Most of the tree bark had already been chiseled off as it
carefully scooted even higher, all four long-clawed legs
wrapped around the branch like a kid shinnying up a tree.
The branch shook with the porcupine's weight and the
breeze as clouds raced past in a blue sky. Then the porcupine
moved its head below the branch as it gnawed on its underside. It had picked the only chestnut oak in the area, one of its
favorite trees.

As I turned to leave, the porcupine slowly descended back-
wards down the branch, again holding on with all four legs,
but only went a short way before resuming its intent feeding.

JANUARY 11. A cold, gray, damp day, but I went outside in late morning and instantly had a good sighting—a
male northern harrier hunting over First Field. Not only did I
see the white patch above his tail, but as he flew directly overhead I had a good look at his white underparts, his wings
tipped with black, and his dark head. I was amazed since I
have never seen harriers here in the winter. They tend to be
occasional autumn and spring migrants.

Later, it cleared and I walked in the bright sunlight with
Bruce. The trail was stitched with grouse tracks, and walking
down Dogwood Knoll, we found a pile of ruffed grouse scat

and an impression of its body in the snow remnants. The grouse must have sat there a long time to excrete so many white and beige, elongated scats in the depression. It probably waited out one of the storms at that spot.

JANUARY 12. The south-facing slopes are losing their snow cover fast so in midmorning I took the Butterfly Loop Trail. I wanted to bask in the sunlight, my back against a large locust tree, and listen to the birds. Mostly, I heard a continual chorus of American crows and black-capped chickadees "fee-beeing" back and forth.

"Bird of the merry heart," nineteenth-century naturalist Bradford Torrey called the black-capped chickadee. Flocks of these popular black, white, and gray birds especially brighten January and February days for me during my walks in the woods.

As early as late August, after their fledglings have dispersed, adults band together in winter foraging and roosting flocks. Each flock, averaging between six and ten birds, includes dispersing young from other areas. They stay within distinct flock ranges of twenty to fifty acres but sometimes combine with another flock to feed over both ranges during the day.

The black-capped chickadee also joins other species, primarily tufted titmice, downy, hairy, and red-bellied woodpeckers, white and red-breasted nuthatches, golden-crowned kinglets, and brown creepers in mixed-species flocks, as I have constantly observed during my winter walks. Such foraging flocks can last for a few minutes, a few weeks, or even a few months.

But at night each chickadee flock roosts in its own range. That is why Susan M. Smith, a faculty member at Mount Holyoke College who has studied black-capped chickadee flocks for many winters, defines a winter flock as "chickadees that typically roost relatively close together throughout the winter" in her book, *The Black-Capped Chickadee, Behavioral Ecology and Natural History*.

Smith, who does her work in Massachusetts, has found that winter flocks of chickadees, which consist mostly of mated pairs, are stable and maintain a rigid hierarchy. Males rank higher than females, older birds higher than younger ones, and larger birds dominate smaller ones. High-ranking chickadees get the richest food sources and safest foraging spots during the winter. If a high-ranking member disappears, it is usually replaced by a winter floater, a young, low-ranking, unpaired chickadee that may travel with up to six other flocks. If a low-ranking bird disappears, it is not replaced by a floater and the gap in the flock remains empty throughout the winter, Smith discovered.

Winter flocks move from food source to food source, using a short, soft, high "tseet" contact call to stay in touch, but on warm, sunny days they tend to scatter more. There are even "dominant wanderers" as Smith calls them, who find food sources far afield, including bird feeders.

Their favorite feeder foods are sunflower seeds and suet. In the wild, chickadees look for the seeds of rushes, conifers, tulip trees, milkweed, goldenrod, ragweed, and wild berries. During October and November, they store seeds and insects in cracks and crevices of tree branches, clusters of conifer needles, and knotholes. Each item is stored separately in a process known as scatter-hoarding.

Researchers have found that black-capped chickadees store as many as a thousand seeds a day, and that they remember what items they stored where for at least a month. That's because their hippocampus, a portion of the brain involved in spatial memory, is unusually large and parts of it regrow every fall when they do most of their food storing.

Chickadees also glean insects, spiders, pupae, and eggs from tree branches and trunks, using specialized leg muscles to hang upside down and obtain food unavailable to less acrobatic bird species.

Other winter survival techniques include having a fresh set

of thick feathers and much denser plumage than other birds their size. They roost singly in small, enclosed areas such as tree cavities, tucking their heads under their feathers and going into regulated hypothermia, an energy-saving process that lowers their body temperature from twelve to fifteen degrees Fahrenheit.

On cold days, chickadees move slowly and when it is windy, they forage lower to the ground or in more sheltered areas. They also feed frenetically when they sense, through a middle-ear receptor that detects small changes in barometric pressure, an approaching storm.

Ever vigilant, the chickadee is almost always the feeder bird that warns, with a high-pitched, thin "zee," of the fast approach of a sharp-shinned hawk or other raptor. Every small bird at our feeders understands the call and freezes. They also understand the "all clear" "dee, dee, dee" call given by the high-ranking chickadee several minutes after the predator disappears.

By late February, chickadee flocks start breaking up into pairs and the male continues to sing the "fee-bee" song that is used to lead flocks in the winter, in territorial disputes during spring and summer, or as a recognition signal. Each pair tries to stake out a territory within the flock's home range. If there is not enough room, the highest-ranking pairs drive out the lowest-ranking ones.

The female usually chooses her mate in the fall when winter flocks are forming. Some remain faithful through several seasons. Smith found that of ninety-four pairs she studied over ten years, seventy-nine stayed together, a better average, I might add, than the 50 percent of human American couples. Of course, a chickadee's average life span is only 2.5 years, which might make it easier to stay together, although a few have lived twelve or more years.

Before the winter flock disperses, each pair explores for potential nest sites and will excavate several cavity sites before deciding on a final one. Then the female builds the nest inside the rotted part of a tree, providing a firm foundation of moss, pine needles, or strips of bark, and lining it with rabbit fur or downy plant fibers.

Nest building takes four days. After a few days of rest, she lays one white egg, streaked and spotted with reddish brown, per day and spends the rest of her time feeding heavily on caterpillars she or her mate find. She lays an average of seven eggs and incubates them twelve or thirteen days. Once they hatch, she broods the naked young while the male does most of the feeding of her and the nestlings. In another twelve days, the nestlings are fully feathered, and the female feeds them as often as the male until they fledge four days later.

Then both parents and fledglings leave the nest site. But the young are completely dependent on their parents for food for several days. Soon they begin catching some of their own meals, and three to four weeks after fledging, the young disperse randomly and sometimes many miles from their natal territory before joining a winter flock.

JANUARY 13. A dismal day, twenty-seven degrees and raining. But in midafternoon Dave stepped out on the guesthouse porch and spotted a least weasel hunting voles near the springhouse. He rushed over for a closer look as the weasel ran through the dried goldenrod stalks to the old well behind the springhouse and disappeared into a vole burrow. Then Dave alerted Mark and while Mark kept an eye on the burrow entrance, Dave ran up to the house to tell me about the weasel.

In the meantime, Mark saw first the vole, then the weasel, zip out of the burrow. The weasel chased the vole up the slope, where we caught a glimpse of it as it disappeared down another vole burrow near the juniper tree outside the bow window. Although we watched that entrance, we didn't see the weasel again. It probably had caught its victim.

Because least weasels are primarily nocturnal and highly secretive, few people have seen them at any time of the year.
The smallest carnivores, least weasels are only slightly larger
than meadow voles, their preferred prey, although they also
like white-footed mice, shrews, insects, ground-nesting birds,
and even carrion. They, in turn, are eaten by barn, barred, and
great horned owls, broad-winged and rough-legged hawks,
foxes, minks, ermines, long-tailed weasels, and house cats.

Most least weasels in northern Pennsylvania turn white in
winter, while in central and western Pennsylvania they are
usually pale brown, as was the one we watched. Their elongated bodies are aptly suited for chasing voles through their
runways, and like other weasel species, they are efficient
killers, grabbing their victims in a death hold by wrapping
their legs around their prey after seizing it by its head and
neck and rapidly biting into the base of its skull, which instantly kills it. Then they eat it, starting with its head and
brain, or else they cache it for later consumption.

A female might bring it to her nest to feed her young. And
those young can be born any time of the year. Least weasels
are the equivalent of bunnies in the weasel family. They can
breed all year, even in midwinter, and the females are able to
have as many as two litters before they are one year old. Usually they have between two and three litters every year. The
young are raised exclusively by the female who, after a thirty-
seven–day pregnancy, gives birth to an average of five wrin-
kled, pink, hairless, and blind pups.

After only four days, they have tripled their weight and are
vocal, emitting high-pitched squeaks. It takes only a couple of
weeks for them to grow their brown coats and be able to eat
solid food, but they don't open their eyes until they are four
weeks old. Then they are weaned and taken on forays by their
mother, who teaches them to hunt. They learn quickly and
are good hunters at seven weeks, but they stay with their
mother until they are three-and-a-half months old.

Least weasels' favorite habitats include brushy areas, open woodlands, old fields, and even mature forests, all of which we have in abundance. Yet today was our first sighting of this amazing creature.

JANUARY 14. Sleeting all night and continuing this morning even as the thermometer stood at ten degrees Fahrenheit. A walk down the hollow in the midst of a heavy sleet storm was like walking through cold, white sand. The birds were out, despite the weather, in a desperate search for food—six northern cardinals, the hermit thrush, and a white-throated sparrow foraged in the frozen streambed above the big pull-off. Tulip tree seeds continued to fall and provide food.

This is the Denver storm all over again—the storm we missed while we were in Denver years ago when ice pellets rolled down the steep slopes and sealed off the hollow road. By this morning, the lowest part of the hollow was already impassable. I stood and watched the ice cascade down, burying the left-hand track in at least a foot of pellets, and moving inexorably across the road. All schools were closed and the storm is supposed to continue for another two days.

Meanwhile, the birds flocked to the feeders and what looked like gray mist was heavy sleet/ice pellets coming down. Sleet, according to the *Encyclopedia Americana* is precipitation in the form of small pellets of clear or translucent ice. Sleet forms by the freezing of raindrops or the refreezing of melted snowflakes falling from a warm layer of air overlying cold air at the earth's surface.

Dave walked down the hollow in late afternoon to check on the sleet accumulation and was rewarded by seeing both the hermit thrush and a porcupine, the latter in a hemlock tree above the big pull-off.

JANUARY 15. Snowing last night. How cozy and secure I felt under piles of blankets as I listened to the silence of snow falling after the hiss and tick of sleet against the windows for so many hours.

Again all schools were closed. The lower part of the road is completely sealed, according to Mark, who took two hours to make it down and back on foot. So I was resigned to waiting for Bruce to clear it with our secondhand bulldozer before venturing outside even after it stopped snowing and blue sky appeared on the horizon at 9:00 A.M. But Dave suggested I go snowshoeing.

To my surprise, the conditions were excellent, not in the least bit slippery—and mine were the only human tracks on the mountain. On the Far Field Road I crossed a turkey track and could see wing marks where it had taken off.

What a joy to be outside on snowshoes again and moving along the old trails on a day lit by the golden white glow of the sun behind a thin cloud layer with here and there patches of blue sky even as occasional flakes sifted down.

Suddenly the sun shone fully, casting long shadows across the snow as the trees creaked and groaned in the wind along Sapsucker Ridge. Clouds raced past; the snow sparkled on First Field, and dried wildflowers and grasses, encased in ice, lent color to the scene—beige, brown, rose-brown—as they glittered in the light. The Norway spruces were bent low with snow, transporting me, for an instant, to the far North.

The boys and Bruce finally broke through the drifts and ice at the bottom of the hollow near dusk, using old-fashioned shovels to chip through the "glaciers" blocking the road, since the bulldozer wouldn't start. Bruce did use the snowblower on the tractor, but it didn't perform as well with ice as the bulldozer would have, so our road is passable, but precarious.

JANUARY 16. An eastern screech-owl called nearby at dawn, but the birds ignored it and flocked to the feeders. Diversity at the feeders is about as high as it can get without the northern irruptive finches—white-breasted and red-breasted nuthatches, mourning doves, black-capped chickadees, tufted titmice, dark-eyed juncos, house finches, American gold-finches, northern cardinals, Carolina wrens, song sparrows, American tree sparrows, white-throated sparrows, and an occasional sharpie—fourteen species in all. I put out some melted chicken fat in a jar lid, covered it with sunflower seed, and tucked it into a corner of the wooden feeder where the Carolina wrens eat. Almost immediately one was poking at the mixture and seemed to be eating it.

Before any birds come to the feeder, while it is still quite dark, a cottontail rabbit siphons up birdseed from the back steps. No doubt it is spending the winter in the abandoned woodchuck hole beneath our front porch where it sleeps during the day. In addition to our birdseed, it eats the buds, tender twigs, and bark of many shrub and tree species in the winter.

Sometime this month or in February cottontails engage in courtship. Snow is no deterrent, as females turn to face advancing males and either jab at them with their front feet or charge them. Dominant males will then dash at the females, and if the females are interested, they jump over the males. One observer—Glenn Wasson—recalls being awakened by his mother in the middle of a cold winter night to look out on a large assemblage of rabbits—"leaping, tumbling and rolling in spumes of flying snow . . ." He thought they were playing games. "The most spectacular game was their version of chicken," he wrote in an essay for *The Christian Science Monitor*, "in which one rabbit would run full speed at a line of sitting rabbits, which would each leap high into the air as the runner passed underneath." Wasson and his family had been treated to the full range of cottontail courtship.

Eventually, after more chasing, the females relent and breed with the dominant males. The rabbits then go their separate ways. After between twenty-five and thirty-five days, the female cottontail gives birth to her first litter in a warm nest she has dug. Measuring about five inches in diameter, the nest is lined with dry grasses and shredded leaves, and finally with soft fur she has plucked from her belly and breast. Exposing her breast allows her three to six young to nurse more easily.

During the daytime, she stays in her form—scratched-out depressions in clumps of grass, under brushpiles, in blackberry thickets, or at the base of suitable trees—away from the nest, returning to it at dawn and dusk so her young can nurse. For sixteen days she tends them, licking them clean after feeding them, carefully closing the nest, and then feeding nearby. The little ones are then on their own, while their mother has as many as three to seven litters in one year, should she live so long. Up to half of her newborns may have young before the year is over. In scientific terms, cottontails are r-selected, which means they have high birth and death rates, providing food for a large number of predators—gray and red foxes, raccoons, fishers, coyotes, bobcats, feral cats, great horned owls, red-tailed, red-shouldered, rough-legged, and marsh hawks, northern goshawks, and American crows.

To escape predators, rabbits use two tactics. Either they "flush," which is a rapid, zigzag run to a familiar travel lane leading to cover, or "slink" by keeping their bodies close to the ground and their ears laid back as they move to cover. Frightened or injured rabbits emit high-pitched screams, bloodcurdling sounds that we sometimes hear at night.

I snowshoed up to the spruce grove and behind one isolated Norway spruce, where Bruce had used his tractor to drag an old apple tree felled by an autumn ice storm, I found all of the apple tree's reachable bark had been chewed off by rabbits. Rabbit tracks radiated from all directions, specifically to the spruce grove itself, which provides both food and thermal cover, and to the Sapsucker Ridge woods.

I continued my snowshoe through a mostly silent forest,

but multiple turkey tracks crisscrossed the trails as well as those of ruffed grouse, squirrels, and foxes. Long blue shadows lay across the snow. The woods were hushed, but the interstate, now opened again to traffic, overwhelmed me with its noise.

It's amazing how different the winter weather can be from year to year. On this date in 1995, it was thirty-seven degrees. I had gone down to the basement to hang up the clothes and noticed a large amphibian foot pressed against the window well above the wash sink. I ran outside and found an enormous American toad up and moving around. It had probably come from a hole in the corner of the window well where it had prematurely emerged in the springlike rains and warmth of the previous days.

It then stood up on its hind legs on tiptoes, stretching its front legs as high as they could go and, pressed against the window, swayed back and forth. So humanlike did the toad appear that I briefly entertained the notion of a frog prince before it went back down into the earth to await spring.

JANUARY 17. The temperature quickly rose from thirty-four to forty-two degrees with a fierce wind that blew the mountain laurel leaves horizontally. A gray squirrel leapt from a nearby tree, panicked by my closeness, and the trees creaked and moaned and sighed as branches scraped against branches and half-fallen trees rubbed against the trees they were caught in. The wind kept changing directions and even in the hollow it blew off and on with a frightening force that kept me alert for falling trees and branches.

Sitting in a side hollow, braced against a sturdy red oak that didn't move in even the hardest gusts, I fended off blowing leaves. There were short periods of calm, like the eye of a hurricane, but the periods of wind were longer. I cringed before the power of Nature.

Once I descended into the depths of the hollow, it was

calm and peaceful, with only an occasional gust of wind. Instantly, I calmed down too and understood why wild creatures are nervous on such days. They can't properly hear or smell danger or food.

Sitting on Waterthrush Bench, I heard the wind on both ridges and looked up Laurel Ridge at fierce whirlwinds of leaves blown so hard into the air that many drifted down to the stream and bench, traveling hundreds of feet before finding another resting place.

By then, the sky was a break-your-heart, absolute blue that only winter winds can produce when they blow away most of the air pollution. As I reached First Field, a wintering red-tailed hawk whistled before it took off from the woods' edge. A common raven called over and over and I finally located it as it rocked in the wind, then streaked along, its wings swept back, dove a couple hundred feet, pulled out of the dive, and repeated the pattern over and over—playing in the wind.

JANUARY 18. Today is another Project Feederwatch day and I spent most of it chasing gray squirrels so the birds have a chance to feed. Woodland gray squirrels are shy creatures that prefer to eat wild foods when they are available. Their favorites are hickory nuts, followed by hazelnuts and white oak acorns, although pecans in the South and black walnuts in the North are also popular. Nut bearing trees have irregular fruiting periods, which prevents nut-eating animals from building up huge populations capable of eating every nut a tree produces. After a particularly good nut bearing year, the trees' supply of stored carbohydrates needed to produce a nut crop are exhausted, so they skip a year.

But the nut predators, such as gray squirrels, have increased because of the previous abundance of food. When the nut crop crashes, so too does the squirrel population. Since most mature deciduous forests have a wide variety of nut bearing trees, it's rare that they all fail at once. Here on our mountain it has happened six times in thirty-two years. At such times squirrel die-offs can be severe.

Malnutrition is the greatest killer of squirrels. Either they die outright from starvation or their weakened condition leads to diseases such as sarcoptic mange. Caused by scabies mites, underfed squirrels lose their fur, a deadly condition in winter when squirrels depend on an outer layer of fur, an underfur for more insulation and skin protection, and a layer of fat to keep warm.

Another strategy males and immature females use to conserve body heat in winter is to sleep together in a drey (nest). The winter dreys are constructed to last. Both sexes build them using twigs on the outside and moss, lichen, fur, feathers, and leaves on the inside. They are waterproof and strong enough to survive heavy winds. But tree cavities are preferred den sites, especially in colder climates.

In both dreys and tree dens, squirrels sleep much of the winter, particularly during severe weather, when they don't come out at all. But on reasonably good days they invade my feeders early in the morning and late in the afternoon, the usual foraging hours for wild squirrels.

Ideally each gray squirrel eats one-and-a-half pounds of nuts every week, or an average of three ounces at each feeding. Unlike red squirrels, gray squirrels do not defend territories or their buried nut supply. Their one- to seven-acre ranges are shared, so whatever one squirrel buries is fair game for its fellow range inhabitants to sniff out, dig up, and eat. Researcher Michael Steele of Wilkes College in Wilkes-Barre, however, has watched gray squirrels pretend to bury nuts at several places before finally doing so, a strategy they may be using to confuse other cache robbers such as chipmunks and blue jays.

At bird feeders gray squirrels prefer black-oil sunflower seeds. But while they eat enormous amounts here in famine years, they eat little or none during good years. They also temporarily invade, as they have the last week or so, when a hard layer of ice makes it difficult for them to dig down to their buried nuts.

JANUARY 19. On this date in 1994 our thermometer was at nineteen degrees below zero, the lowest it has ever been here. But today it is twenty-eight degrees above zero, heavily overcast and lightly snowing.

A walk in the snow along Greenbrier and Ten Springs trails yielded only a northern cardinal, white-breasted nuthatches, tufted titmice, and black-capped chickadees. But as I neared the end of Ten Springs Trail, five American crows cawed loudly in the uncut forest ahead. I sat down in the heavily wooded hollow and, looking uphill, I quickly spotted the shape of a great horned owl sitting on the end of a tree branch. It looked as if it were unbalanced, an extension of a branch that leaned but, in fact, the branch had a projection against which the owl was braced.

I sat and watched for fifteen minutes as two crows remained nearby, occasionally emitting low noises of approbation in their throats. Then I spotted a second, slightly larger owl sitting tightly against the trunk. No doubt it was a pair— the larger female against the trunk, the smaller male out on the limb. I'd never seen a pair before so I continued watching them until I was too stiff and cold to remain any longer. As I stood up, stretched my cramped legs, and continued my walk, the male, who had had his head turned away, finally looked down at me. The female never moved. With both owls, it had been their ear tufts that had made it possible for me to first pick them out despite their frozen stance and excellent camouflage.

January is great horned owl month here. Not only are their hoots the quintessential signature of long, silent, moonlit winter nights, they also are more visible in the daytime. During the rest of the year I may have an occasional glimpse of one as it flies from a roost, but my best sightings always occur in January. Another notable one happened back on January 18, 1993, on a windy, twenty-degree, sunny morning.

Again scolding crows tipped me off and I tracked them to Margaret's Woods, but I could not see what they were fussing over. Still they persisted, so I finally walked toward them up the Steiner/Scott Trail that leads to the top of Sapsucker Ridge, a trail, incidentally, that we named for two now-deceased hunters who used to hunt on Margaret's land—Cloyd Scott and Harry Steiner.

Halfway along the trail I spotted a large, beige-colored blob on the branch of a tree swathed in grapevines. To my delight that "blob" was a great horned owl bathed in sunlight and not inclined to fly. I sat below it, my back against a tree, and watched as a pair of black-capped chickadees flew in close to scold the owl. Then a tufted titmouse joined the chickadees as they called a couple of feet from the owl's head. It blinked its eyes open to look and the birds flew off, so it closed its eyes again.

After a few moments, it again opened its eyes and slowly turned its head to the side. The eye toward me watched as I stood up. After I took a couple of steps, it flew off.

I was amazed at how perfectly it had blended into the tree branch and grapevines. Without my binoculars I never would have distinguished the owl with its perfect camouflage. I thanked the crows, which, for once, had had something to crow about.

According to biologist Bernd Heinrich, who studied mobbing behavior while raising an orphaned great horned owl, birds that are permanent residents of an area, such as the titmouse and chickadees I noticed, use mobbing to encourage owls to move on. He further hypothesizes that "since crows have conspicuous roosts to which they return each night, the move on hypothesis should apply especially strongly to them. And, indeed, the vigor of the crows' mobbing in winter is surpassed by few other birds, even in spring." Furthermore, after analyzing owl pellets in his woods, he discovered that crows were the primary prey of great horned owls there.

Bruce and I went out after supper, while snow still fell, to walk in First Field. Less than two inches had fallen, but it was enough to blanket the earth in a new suit of white. We were serenaded by great horned owls on Laurel Ridge—the female a pleading, warbling contralto; the male a grumbling bass interrupting her song—love-duetting in the midst of a snowstorm.

Researcher John T. Emlen had sharp enough ears to detect grunting noises the male makes before he hoots and in between hooting, to stimulate the female to hoot. In turn, the female hooting stimulates the male to continue hooting. All of this is part of the annual courtship ritual between birds that mate for life.

Solitary hooting by males is done to advertise and defend their territories, which they hold on to throughout the year. Most great horned owls remain in the same area where they hatched, unless food is scarce. Then the young may move on, as far as 837 miles in one documented case. But established pairs may not breed if food is scarce; evidently faithfulness to their territory is stronger than the urge to breed.

When there is enough food, courtship takes place mostly in early January and February evenings, and includes calling, displaying, mate-feeding, and allo-preening. According to researchers, it goes something like this: First the male approaches the female by hooting and landing on perches close to her. She may answer him if she is interested. He then performs one or more of a repertoire of displays such as fluffing his body feathers, partly spreading his wings and bowing, walking and hopping on the ground, and throwing his head back and repeatedly snapping his bill. If she doesn't respond by fluffing her feathers or snapping her bill at him—two signals that she is not interested—he then sits on the same perch with her, gradually sidling closer, until they preen each other by pecking at the feathers around their mate's bill and/or head, and may also at this juncture emit a variety of barks, screams, whistles, and hoots. After this no-doubt highly romantic exchange, both hop and bow, and occasionally the male brings in food for them before they mate. After mating, the pair often roosts together during the day, like the pair I observed today.

Great horned owls never build their own nests. Instead, they occupy old nests of red-tailed hawks, American crows, common ravens, great blue herons, or squirrels. Last winter I examined dozens of old squirrel nests in search of nesting great horned owls but never found any. Because they occupy their nests and lay their one to four white eggs from mid-February to early March in Pennsylvania, deserted nests are easy for the owls to preempt. The female primarily incubates the eggs for twenty-eight to thirty days and is occasionally relieved by the male, but once the eggs hatch, the female keeps the young warm while the male provides food. The young hatch over several days, in the order in which the eggs were laid. If food is abundant, all of them survive and thrive. If it isn't, the largest owlet is fed first and competition is fierce. Often the youngest (and weakest) die.

At two weeks of age their eyes are open; at three, brooding stops; at four to five weeks they can move about the nest; and at six to eight weeks they leave the nest, perching on nearby branches where their parents continue to feed them. When they are nine to ten weeks old, they attempt to fly, and gradually, after ten more weeks, they have learned to fly and have been taught by their parents to hunt well enough to disperse, although half of all young do not survive their first year.

Opportunistic feeders, great horned owls tend to eat whatever is available, including rabbits, foxes, porcupines, skunks, other owls, hawks, crows, feral cats, mice, and rats. They fish in water up to their stomachs for fish, turtles, crayfish, and frogs. Heinrich's captive Bubo distinguished between wood frogs, which it relished, and bullfrogs, which it disliked.

Years ago, when we raised Muscovy ducks, we arrived home after sunset and watched a great horned owl trying to fly off with our alpha male Muscovy named Big John. Big John flapped valiantly as the owl aborted a couple liftoffs before giving up.

A food study done in Pennsylvania from 1965 until 1986 by biologists Wink, Senner, and Goodrich analyzed owl pellets from seventeen counties and found that although cottontail rabbits were a favorite prey item of great horned owls (15 percent), ruffed grouse as prey ranged from 9 percent in northwestern Pennsylvania to 4 percent in southeast Pennsylvania, and ring-necked pheasants 3 percent. The overwhelming favorite prey item was the Norway rat (24 percent). But if the diet analysis was based on the weight of the prey, opossums would constitute 33 percent, rabbits 28 percent, and Norway rats 12 percent.

The rats are an indication of how popular farm habitat is with great horned owls. According to a study carried out by Yahner and Morrell from 1986 to 1989 in south-central Pennsylvania, great horned owls hunted most extensively in agricultural areas and adjoining woodlands. In fact, the more fragmented the landscape, the better the feeding opportunities for great horned owls.

Because they will eat almost anything and live almost anywhere—from Arctic Canada and Alaska to the southern tip of South America—great horned owls will survive and thrive long after specialists have gone extinct. That's good news for those of us whose winter nights are enhanced by their antiphonal hooting.

JANUARY 20. Winter woodpeckers keep the forest alive, especially the "peek, peek" calls of downy woodpeckers. And every day a male downy visits our feeder. Sometimes a female comes in too. I often see downies in mixed-species flocks in fall and winter, especially with black-capped chickadees or tufted titmice. Both species give alarm calls when predators appear, enabling downies to concentrate on foraging, knowing that the so-called "sentinel species" are on the alert.

The dapper black and white woodpeckers, the male with a red patch on the back of his head, have smaller bills than their look-alike larger relatives, the hairy woodpeckers. Both its common name "downy" and species name "pubescens" refer to the soft white feathers of the white stripe on its lower back, which early eighteenth-century English naturalist Mark Catesby contrasted to the more hairlike feathers on the hairy woodpecker. Every downy also has a unique pattern on the back of its head, which is useful for both researchers and other downies in distinguishing one bird from another.

Downies are the most common woodpeckers in eastern North America and come to bird feeders year-round. Most folks feed them suet, but at our feeders they are content with sunflower seed and corn. That's probably because there is plenty of natural food for them to eat here. They particularly like to peck and probe for prey items in black, red, and white oaks, but like great horned owls, they are opportunistic, feeding on trees suffering from insect infestations. Seventy-six percent of their food is animal, including forty-five insect and other arthropod species. The rest consists of at least twenty plant species. They are especially known for their fondness (and replanting) of poison ivy berries and have been observed sipping sap from yellow-bellied sapsucker wells in the spring.

In the fall they excavate several roosting cavities that are situated away from the prevailing wind so that they will provide snug, solitary refuges for the downies during long, cold winter nights. And it's every downy for itself until late winter when they begin courting, mostly by drumming, chasing, and calling. Drumming is used not only to attract mates but to establish and defend territories. Males and females also drum at the same time or alternately in what is called "duet drumming," which may be a signal that either sex wants to copulate. Downies have favorite drumming trees as well as distinctive one- or two-second bursts of loud and rapid drumming.

Once downies pair up, they move around together in search of a suitable nesting tree. Most nest trees and snags are twenty to thirty feet tall and ten to fifteen inches in diameter. The nest hole itself is usually fifteen to twenty feet from the ground and goes down eight to ten inches. Both sexes excavate the nest cavity, although the male does most of the building. He also does more than his share of incubating the three to eight white eggs, including all the night shifts.

After twelve days the eggs hatch and again the male takes the lead. Not only does he help to brood the nestlings during the day, but he is solely responsible for night brooding and the removal of fecal sacs.

Both sexes feed the young, foraging within three hundred to four hundred feet of the nest cavity. After twenty to twenty-five days downy nestlings fledge, but they are still dependent on their parents for food for several more weeks.

Black rat snakes are major predators on eggs and nestlings. So are all squirrel species. Adult downies are sometimes taken by American kestrels, sharp-shinned and Cooper's hawks, and eastern screech-owls. Most downies die shortly after fledging because they are inexperienced with predators, flying, and food gathering.

Although downies prefer to live in relatively young, deciduous forests with a low canopy in bottomlands, they seem to be content with our mountaintop habitat, judging by those I see on my daily walks in the winter. After all, they only need live and dead trees, which we have in abundance, to feed on throughout the year.

JANUARY 21. A red sunrise lit Sapsucker Ridge after yesterday's four inches of fresh snow and it was nineteen degrees and clear. Birds flocked to the feeders and the back porch where I had spread seed—ten American tree sparrows, fifteen dark-eyed juncos, four mourning doves. Then, at 9:30, I heard a familiar "thump" at the bay window. There was the sharpie, sitting on its usual spot below the back porch, plucking a junco, gray feathers flying, until it reached the real meat. Despite what seems to be a preference by sharpies for juncos, at least at our feeders, Project Feederwatch reports every year that juncos are, by far, the most abundant species at feeders nationwide.

Mine were the first human tracks as I set out midmorning up Guesthouse Trail, down Laurel Ridge Trail, and on to the Far Field Road. I found many deer and squirrel tracks and finally tracks of a fox that had leaped over the tree stump seat near the beginning of the road and had been joined shortly by another fox that had come down the road embankment. They had paused to leave drops of urine on one clump of fallen leaves before one headed below the road and the other went toward the Far Field. Already I have been smelling the faint skunky odor around the spruce grove that signals red fox courtship.

JANUARY 22. A red-tailed hawk screamed repeatedly as I left the house, but I couldn't locate it. All the trees and branches still held snow cover along the Far Field Road because yesterday's wind did not reach this sheltered area.

I followed two sets of coyote tracks all over the Far Field, where red foxes had always lived. They were classic coyote tracks, between two and two-and-a-half inches long with distinctive toenail marks and heel pads. I saw where the coyotes had sniffed at the bases of trees and poked their heads under blowdowns. Neither burrow I knew of had tracks going in and out, but above Pennyroyal Trail, the tracks converged on a hole beneath an uprooted tree.

A deer had recently gnawed on the bark of a small striped

maple tree near Sapsucker Ridge Trail. Deer tracks wound along the edge of the spruce grove while porcupine tracks penetrated its interior.

Later, as we were finishing lunch in the kitchen, I looked out across First Field at Sapsucker Ridge and, against the revealing backdrop of white, I spotted what looked like a red fox trotting down the ridge.

"Get my binoculars, quick!" I shouted to Bruce and Dave as I kept my eyes glued on the moving animal. Through my binoculars I could see that it was a large red fox. Unlike most red foxes, however, it had black along its back and tail as well as on its legs, which made it a cross red fox. Because of its size, I assumed it was a male.

As we watched from the veranda, it reached the edge of the field and sat down next to a large tree where it groomed its tail and chest before moving back up into the thickets. After a few minutes, it reappeared, trotting along the side of the hill, behind a fallen log, and into thick brush near the powerline right-of-way.

Although red foxes are primarily nocturnal hunters, in the winter they are more likely to hunt during the day in addition to the night because prey is harder to catch. Excited by the sighting, I waited nearly an hour to give it a head start before setting out to track it.

Following up the edge of the woods along the powerline right-of-way, I found where it had crossed about halfway up Sapsucker Ridge. On the other side of the right-of-way it had sniffed around an uprooted tree before continuing into the woods and then back down into the grapevines and American bittersweet hanging from the trees along the edge of First Field. From there it had meandered along the field border for a couple hundred feet before turning back up the hillside into the woods.

Its tracks were nearly as large as those I had seen at the Far Field, but because red foxes have a lot of hair growing on the bottoms of their feet, they were blurry and the toenail marks were not visible. But I could see its distinctive heel pads, which were shaped like an inverted V.

Halfway up the slope it turned left and continued straight ahead for several hundred feet before swerving right and going to the top of the ridge. It headed along the crest toward the vernal ponds, but then veered toward the top edge of First Field, staying well within the woods, even at the corner of the field beyond the spruce grove. There it had paused to inspect an uprooted tree with a nice cavity below.

Finally, it turned left, putting its tracks precisely into the oval-shaped grooves made by porcupines that led into the spruce grove. I lost the trail in a maze of deer tracks—it was as if the fox had simply vanished. Or had it realized that I was tracking it and deliberately thrown me off the trail?

Whatever had happened, I appreciated the chance to track a red fox and learn a little about how one moves through its habitat.

JANUARY 23. Our world was encased in ice after a night of freezing rain. I could hear the smashing down of a tree as I went out to feed the birds. Bruce hosed down the juniper tree with hot water because it was bent over at an alarming angle from the weight of ice. I was afraid it would snap and the dark-eyed juncos would lose their nightly refuge.

At noon, the sun flashed in and out, and a wind picked up, flinging ice off the weighed-down trees, telephone and electric lines. The ice was melting so fast that it sounded like a hard rain, and I was eager to get outside before the glittering beauty was gone.

The road to the Far Field tunneled through sprays of ice-covered, bowed-over black birch saplings. One had been recently gnawed around its base by a porcupine. Fresh turkey tracks meandered along the Far Field Road. They had scratched up an area that dark-eyed juncos were making use of. One ruffed grouse flushed along the Laurel Ridge Trail, and a hairy woodpecker was abroad and working on tree trunks.

As I entered the Far Field, a red-tailed hawk screamed and took off. It kept calling from a distance as I sat against my favorite log. The sky was a jumble of silver-gray clouds with a small band of gold on the western horizon and occasional patches of blue sky. Both ice and water fell from the trees. Along the inside of one ice-encased tree trunk, I watched small drops of water running down that looked like one-inch–long, skinny worms. I quickly soaked my gloves while removing ice castings of broken hickory branches, laurel leaves with two-inch–long, frozen-tip drip lines, black birch and witch hazel cones, and grass stems. I am endlessly fascinated by these ephemeral ice sculptures of nature.

JANUARY 24. Winter is a seesaw here, erratically moving up and down from almost-spring to definitely-winter in a few hours. Most of the snow was gone this morning after a warm rain overnight, and it looked and smelled like late March.

Squirrels were courting as I sat on Coyote Bench. They paid no attention to me as four males chased a female, but one seemed to be the alpha male and fended off the others. Once she turned and faced him at the end of a branch, he retreated. Then she found refuge in a tree hole and nipped at any squirrels that tried to get in. She kept poking her head out as if watching the fun, but apparently, according to Michael A. Steele and John L. Koprowski, in their book *North American Tree Squirrels*, her actions were designed to send the males into a greater frenzy, which they did.

The alpha male was busy defending the hole from above and below as the other squirrels tried to breach his defense. When he drove off all but one other competitor, she emerged, ran down the tree and raced off into Roseberry

Hollow. A Johnny-come-lately rushed up the tree she had been in, following his nose, and making the low, grunting noises of a questing male, put his head inside the hole, sniffed around the outside, then ran down the tree again.

At that point the female, pursued by five males, streaked up from the ravine and across the road. The last I saw her, she had six males, more or less, on her tail as they disappeared. This courtship day probably started shortly after dawn and will continue for eight hours. In that time she will mate three or four times with the same alpha male or maybe with one of the others.

I walked on to the Far Field where the dried, orange sprays of little bluestem glowed among the brownish-purple dried goldenrod and asters, like a row of burning bushes. Later, a grove of small, American beech trees, dangling their almost-white leaves, similarly startled me into reverence.

Today the sun returned to Barrow, Alaska. That means our days too will lengthen. To me winter is both its own season and a foreshadowing of the long, light-filled days of spring.

JANUARY 25. Cold and windy, with a new skim of snow on the ground. I set out for my usual morning walk, keeping to the shelter of the woods until I started down the Far Field Road.

That was when I discovered enormous, roundish tracks with well-formed toes coming up the valley from Roseberry Hollow, crossing the road, and heading across First Field to the spruce grove. Since it had stopped snowing at dawn, the tracks had to be fresh. I also knew, even though I had never seen such tracks in the middle of winter before, that they were those of a bear. What, I wondered, was a bear doing out at this time? Why wasn't it hibernating?

With only a little trepidation, I followed the tracks into the dense spruces, certain I would find a bear bedded down there and knowing that I would see it only at the last minute. Instead, the tracks continued through the spruce grove, across the top of First Field and into Sapsucker Ridge woods. Eventually, after wandering up and down the hillside, they led straight to an uprooted tree with a smallish hole beneath, an ideal den site, I thought. I circled the area cautiously, but once again I was fooled. Those tracks continued on the other side of the root ball.

On I went, at first following them as they continued through Sapsucker Woods halfway between the ridge top and First Field. Then they went up and down the ridge, across the powerline right-of-way and along the woods' edge above First Field. Had we looked out our kitchen window at sunrise, we might have seen the bear.

Sapsucker Ridge has a southern exposure, so as the morning warmed, the snow began to melt. This made tracking more difficult because there were portions of open ground between the snow patches. But once the bear entered a snowy patch of woods, its tracks were again easy to follow.

Several times the bear climbed down the trunks of fallen trees, and it seemed to be heading for the large crawl space under Margaret's derelict house, a place that had struck me more than once as an ideal bear denning spot. But as the tracks neared the house they veered once again toward the ridge and the growing-up-to-brush clear-cut area.

By then I had been tracking the bear for a couple of hours, so I took time out to go home, eat a snack, and persuade Bruce to come along armed with his camera. The tracks went halfway up the ridge, over, under, and along fallen debris until they reached what again looked to me like an ideal den site—a huge hollow log swathed in grapevines with a large hole beneath.

Yet we could find no tracks into the hole and, in fact, lost the tracks completely on the mostly bare ridge. In desperation we climbed to the top of the mountain where there was more snow, hoping to pick up the tracks as they crossed the logging road, but we found nothing. Exhausted by then, we decided to give up the search and go home for lunch.

Dave listened carefully to our tale of bear-tracking over the lunch table. Then he went out to retrace our steps. He picked up the faint remains of our tracks in the clear-cut and realized that the bear had doubled back on its tracks. Discovering the bear's trail farther down the ridge, he finally tracked it to a ravine still filled with logging debris. There he found a full-size bear sleeping on top of the snow. The bear leaped up and ran off just as Dave approached.

Because in late January most adult female black bears are either having cubs or denning with their cubs from the year before, the bear we had tracked was probably a male. Had he been disturbed in a valley den site and come up the mountain to look for a better place to sleep out the winter, or was he merely restless?

What had induced him to venture out on such a cold, windy day? Certainly not food. Black bears neither eat nor defecate during the winter, although I did find three places in the snow where he had been digging. On the other hand, I found no droppings.

Later, I checked back in my journal and found a reference to winter bear tracks on another cold clear day—January 28, 1995. I had gone up on Sapsucker Ridge to look at the four frozen vernal ponds and found a day-or-two-old set of bear tracks on the second pond. Those tracks had continued on to the third pond where the bear had crashed through in one place and slid on another. Finally the tracks had crossed the largest of the ponds where it had again crashed through, this time along the edge, before heading up the final slope. While the ponds had still supported a skim of snow that showed tracks, the woods were bare so I could not follow them any farther.

According to the books, all black bears should be sleeping away the winter in their dens by early winter. Yet at least two bears were wandering our mountain in late January. Obviously, not all bears remain dormant for the entire winter. Maybe they, like me, were eager for signs of spring.

JANUARY 26. Or maybe yesterday's bear had sensed that the January thaw was imminent. It was forty-three degrees at dawn, clear and breezy. Three eastern bluebirds sat on our electric line. American robins "tut-tutted" and tufted titmice "peter-petered" from Sapsucker Ridge. I flushed six ruffed grouse from under a thicket of grapevines along Big Tree Trail. They favored the sunny slopes, and so did I. Sun burned into me as I sunbathed against the Far Field road bank—the warmest spot in the winter, the coolest in the summer. Farther along the road at Coyote Bench, it was warm but silent. The squirrel orgy was over.

In midafternoon, it was fifty-eight degrees in the sun and the honeybees were flying in a midwinter cleansing flight, which is, in reality, a chance to defecate since they never do so inside their hives. One landed on me as I sat on the veranda reading and examined a ladybug beetle on my sweater before buzzing off.

Dave's herb garden glowed with a patch of twinkly-blue field speedwell *(Veronica agrestis)* blossoms. Although it is another tough alien, like dandelion, according to *The Plants of Pennsylvania* its regular blooming time is April through July. Still, any flower that appears at this time of year (and none ever has before) is welcome no matter what its origin may be.

JANUARY 27. The thaw continues. Forty degrees at dawn and the temperature rising quickly. An American robin called in the yard at first light, and a Carolina wren sang on and on from a black walnut tree in the yard shortly after 9:00 A.M.

The sun shone into the forest, lighting up every tree trunk, the large and the small, the crooked and the straight. A mourning dove cooed its sorrowful song on the top of Laurel Ridge.

Walking in the leaf detritus beside Big Tree Trail was a bright green stinkbug that looked perfectly chipper. I carefully edged around it, not wishing to crush such an adventurous, brilliantly colored insect abroad on a winter's day.

Sixty-eight degrees in the sun by early afternoon. House finches sang. Two pairs of eastern bluebirds appeared, one on the electric wire, the other in the yard on a black walnut tree branch. All four were diving after insects.

Ladybug beetles were everywhere—inside and outside. This newest alien invader has been stretching our toleration limits ever since October of 1993, when they first appeared on our veranda in the warm autumn sun.

At first I was delighted with the rotund, dark-spotted, orange beetles. All the more ladybug beetles to eat aphids—a single ladybug larva eats as many as three hundred aphids in the two weeks before it pupates. And aphids attack many trees, including apples, peaches, plums, pines, maples, oaks, and tulip poplars, all of which grow in our yard and surrounding woods.

Then, as more and more appeared each October and overwintered in our home, I tried to identify the species. This was not an easy task because there are more than four thousand ladybug species worldwide, including 475 in North America alone. Until October of 1995 I was clueless. But the same beetles that had swarmed here for three years also swarmed in far greater numbers in selected homes and businesses all over the northeastern United States. Those affected begged entomologists to do something.

Articles appeared in newspapers and identified the species as *Harmonia axyridis*, an Asian species that was introduced repeatedly by entomologists to control tree-dwelling aphids. Nicknamed the "Halloween Ladybug" because of its orange color and the time of its swarming, the Entomological Society of America has named it the "multicolored Asian lady beetle" because of its variable color scheme. Although most are yellow-orange, some are brick-red, and a few are even black with orange spots. Those ordinarily black spots not only vary in color but in numbers, ranging from few or no spots to as many as twenty. And, of course, they aren't true bugs, but beetles.

The ladybugs need cool hibernation places, and while most of them hibernate outdoors beneath leaf litter, under the loose bark of trees, or in clumps of grass, others prefer to move indoors. Old white houses in wooded areas are favorite hibernating spots, hence the popularity of our home and the guesthouse. Supposedly, those trapped in our warm living quarters will not survive, the experts say. Maybe not, but they are lively enough, flying and crawling over windows, sinks, and my houseplants. Sometimes several hundred appear on our bay window, and in Bruce's warm study they swarm on every sunny day. In the guesthouse, they appear by the thousands and Dave has resigned himself to living with them throughout the winter, only occasionally sweeping up windrows of the dead bodies in his living room and dining room. I, on the other hand, vacuum them up by the hundreds every warm day, but they are quickly replaced by others.

At first we thought they were cute, and I whimsically compared them to opossums because they roll over on their round, hard wing covers and play dead for several minutes after they are touched. Then they wriggle their legs, hoist themselves over on their stomachs and resume flying or crawling to wherever they are going.

Scientists are not happy about this population explosion, because they are afraid the multicolored Asian lady beetle will wipe out some of our native species of ladybugs. My own patience has worn thin over the past several winters as their numbers have burgeoned.

"Ladybug, ladybug, fly away home" has become more

than an old nursery rhyme to us as we fervently wish they would fly back to Asia where they came from. Instead, we watch this alien species take over our home every winter, especially on these warm, January thaw days, and give us more communion with the insect world than we would like.

JANUARY 28. Forty-eight degrees this morning and a male downy woodpecker came to the tube feeder in an effort to avoid the honeybees swarming over the wooden feeder. The bees are convinced that our seed chaff is an early source of pollen. The red-breasted nuthatch also preferred the tube feeder. Other birds fed on the steps.

It was still warm as I set out on my walk. I wore spring cotton pants and a spring jacket and, like the ups and downs of aging, I was also enjoying this odd, silent spring in January. Even on the recovering clear-cut, where I walked along Bird Count Trail, I found something to admire—a flock of cedar waxwings, ruffed grouse, and American tree sparrows amid the blanket of alien and native species gone amok—hay-scented fern, foxtail grass, privet, Japanese barberry, ailanthus, and multiflora rose.

Where would we start to manage this altered, devastated land for the "right" species? By poisoning all the aliens and natives that have spread because of humans' bad stewardship and then planting and fencing desirable trees? That is what forest stewardship proponents recommend. But so far we have kept poisons off our land. We watch and wait and hope that some good will emerge from the slaughter of every upright tree by our former lumberman neighbor.

Our management of the earth and its creatures solely for the monetary benefit and recreation of humanity has backfired horribly. So the managers are managing the managers now that nongame species are gaining some charisma in the eyes of the nonhunting public. Instead of managing for optimum game species, some managers are suggesting that we manage for nongame species as well, principally by cutting back the deer herd. I fear that natural processes are more complicated than that and that once the deer numbers are lower, other, unanticipated problems will arise as they always do when humans try to get the best of nature.

Along Greenbrier Trail, a flock of white-throated sparrows was eating privet berries. Because most of the native trees were cut by the lumberman and the native shrubs eaten by deer, our native birds are busy using invasive species for food and shelter and, in the process, spreading their seeds. If we poison the thousands of invasive shrubs and tree saplings on this clear-cut, where will the wintering birds seek food and shelter?

Other solutions are proposed by visionary environmentalists. Dave was inspired, the other day, by the launching of a Wildlands Project in Pennsylvania. The hope is to rewild the Appalachians from New Brunswick to Georgia with the eastern cougar as the keystone species. These cockeyed optimists believe that humans can live in harmony with wildness. Their plan is to use large blocks of public lands and purchase key private land to provide corridors through populated areas so that we can bring back the large carnivores that need thousands of acres, as well as the other, smaller species that are greatly diminished because of habitat loss such as bog turtles and wild orchids, and thereby have a healthy ecosystem with all the missing parts in place. No more overabundance of deer eating the forests down because the cougars are gone, for instance. Instead of invasive species in our clear-cut, trees would have a chance to grow and crowd them out. What a vision for the twenty-first century! It seems an impossible task, a momentous undertaking, but Dave is determined to be part of it.

For all its invasive species, the clear-cut was filled with wildlife this warm day. A fox squirrel ran downslope and I sat at the base of the double white oak tree on Dogwood Knoll, listening to a bevy of calling birds as if the sun, newly emerged from a hazy sky, had invigorated them.

I walked along trails that are never open in winter, including the Ten Springs Extension. Only in the deepest, shadiest pockets of the hollow beside the stream were there any patches of ice or snow left. Cool air streamed up from the hollow as I descended the trail through the intact forest, but it was balmy enough to have melted almost every vestige of ice and snow on the road.

Chipmunks, lured from their semihibernation in their underground burrows by the unseasonable warmth, foraged above the stream, and their calls followed me as I headed up the hollow road.

The day remained almost completely overcast, but warm, the sun a dim orb in a mostly white sky—a winter sky and winter woods, despite the warmth. And no rush of melt water has filled the ditches and stream as it usually does in spring.

After vacuuming up the plague of ladybug beetles in the sitting room and kitchen, I spent another relaxing afternoon on the veranda, watching a brown creeper feed, spiraling up a black walnut tree trunk while white-breasted nuthatches moved forthrightly down head first, each species finding different prey on the same tree.

In the evening Bruce and I stood out on the lawn and listened to earthworms rustling in the leaves as they pulled pieces down into the thawed earth where they could process them into soil. Never before have we heard this any earlier than March.

JANUARY 29. It was fourteen degrees early this morning: the thaw is over. I went outside at 9:00 A.M and once again heard the scream of a red-tailed hawk. Once again I looked around and didn't see one. This has been happening nearly every day. How can my eyes be so bad that I can't spot a red-tail? Then, suddenly, a blue jay flew into the yard, landed on a small tree not far from the feeder, and rendered a perfect red-tail scream, not once, but three times, and pointed in my direction as if to say, "You're not crazy or blind. It's me."

Dave came up from the guesthouse and said that at least one jay has been working on perfecting the red-tail scream since August. "And I can't tell it from a true red-tail scream anymore," he said.

I agreed. So now, unless I see a red-tail, I will assume that the scream is a jay imitation. This is only the second year we have had wintering blue jays, which is why I was so easily fooled. Why do we have these birds now after at least twenty-nine years without them?

When I set out on my walk, I heard a loud, glassy thump as I passed the garage. A blue jay was throwing itself against the window in an effort to escape. I tried to encourage it to fly out of the garage's open doors, but instead it flew high in the rafters and refused to move so I left it to its own devices.

Nature repeated itself and so did I. I carefully removed my sunglasses and adjusted my binoculars as I followed the trail around the edge of the spruces. The large bird I have been trying to identify erupted from the spruce ahead and flew across First Field, landing in a large tree in the woods beside First Field Trail. I thought I had my eyes fixed on the spot, but I could not locate the bird with my binoculars as I walked toward the trees. Foiled again! In early afternoon, I heard screaming and ran outside to look around.

"Kak, kak, kak!" It was a male American kestrel. Where did he come from? I've never seen one here in the middle of winter although I know that a few do winter in the valley. No wonder the number of small birds at the feeders keeps dwindling. Too many raptors are around!

At 5:00 P.M. I walked back up to the spruces and sat among them, watching the orange light of sunset fade into darkness, listening first to the "chips" of dozens of dark-eyed juncos as they settled into the spruces for the night. Once they were quiet, the whistling wings of mourning doves broke the silence as they jockeyed for position in the dense evergreen trees. Neither bird species seemed deterred by whatever large bird is hanging out there.

I walked back as the first stars lit the sky and was serenaded by courting great horned owls. The peace of the spruces and the spectacular orange and golden sky made me profoundly grateful for such moments of grace.

JANUARY 30. A Carolina wren foraged below the feeder area this morning. Dave says two of them caroled back and forth in the yard shortly after daybreak. Both must have been males since females never sing. Once again there are plenty of Carolina wrens on the mountain.

It was on this date, in 1998, that the Carolina wrens returned. I was walking on Ten Springs Trail and heard birds calling, so I pished and the expected dark-eyed juncos emerged from the underbrush, along with the unexpected: two Carolina wrens! What a joy it was to see and hear them. They had been gone since February 19, 1993, when they had disappeared after finding their way into our basement through mouse tunnels. One even spent the bitter, four-degrees-below-zero night of February 18 perched on a hanging pot in our kitchen. The following day I released the bird. She was happily greeted with a burst of song by her mate who had spent the night outside. That was the last I saw or heard them. Since then I had sorely missed what I consider to be the most personable of all the songbirds.

Because their replacements were a mile down the mountain from our home, I did not expect them back any time soon. But by March 1 they were singing in Dogwood Hollow, an advance of several hundred feet. During the rest of the spring, though, I heard them only in Clearcut Hollow, a brushy, wet place crowded with sapling striped maples and black locusts, so I assumed they were raising a family in the vicinity.

Then on June 8, a Carolina wren once again sang in Clearcut Hollow. But as I proceeded on Ten Springs Trail into the uncut forest and wound my way down to our road, a whole family scolded and sang—a set of fledglings foraging with their parents a mile from our home.

Unlike most wren species, Carolina wrens are monogamous and the males are as busy taking care of the young as the females. They can have as many as three broods in a season, but here in Pennsylvania, in the northern part of their range, two broods are probably the limit. They like a wide range of habitats, from brushy clear-cuts (our Dogwood and Clearcut Hollows) to hemlock-and-rhododendron–choked ravines (our uncut forest near the stream).

Only males sing the joyful "tea-kettle" song, which they have nearly perfected by the age of forty days, when they disperse from their parents' territory in search of their own turf. There they sing all year long, even in the teeth of blizzards, as I had discovered, defending the territory they occupy throughout the year. Researchers claim that their singing is dependent on a good supply of food and that they don't sing during abnormally cold and snowy weather, but our 1993 male had not read the books. Researchers did discover, however, that one captive male given unlimited food sang nearly three thousand songs in one day!

Carolina wren fledglings are known to disperse as far as a fifth of a mile, so where did the Carolina wren singing in our yard on June 29 come from? Hadn't he read the books either? Had he dispersed from the singing wren family in the hollow or had he come from somewhere else on the mountain? I found that difficult to believe because I cover the area on a network of trails daily and had heard no other Carolina wrens.

By July there even seemed to be more than one singing. Around our house Carolina wrens chortled from dawn until dusk. In August we clearly identified one male who sang in the vicinity of our front porch at dusk just before retiring for the night in an abandoned eastern phoebe nest built on top of the corner column of the veranda. Every morning he was up at first light to greet the day and was the only singer breaking the dawn silence.

Bruce, who was as happy about the return of the Carolina wrens as I was, whistled their song and then whistled the opening theme of the first movement of Vivaldi's Concerto in D, which sounds very much like the Carolina wrens' song. Yet, unlike our winter wren, which is the same species as Europe's wren, our Carolina wren is an American original. Vivaldi never heard the song of the Carolina wren. But we are once again privileged and grateful to be hearing its jubilant notes every day of the year, especially in winter, when he (and his mate) are only occasionally joined by the wintering song sparrow, and, in warm spells from now on, by black-capped chickadees and tufted titmice.

JANUARY 31. Nothing is more surprising on a cold winter day than the sight of an insect crawling over the snow or fluttering in the frigid air. Most insects are cold-blooded so their body temperature is the same as their surroundings and cold weather usually stops their activity. They enter a period of rest called "diapause" and remain inactive until spring.

These overwintering insects—whether in adult, larva, pupa, or egg stage—produce their own antifreeze or glycerol compounds, which enable them to survive subfreezing temperatures. But first they add fat before diapause. Then they drastically reduce their body liquids and "supercool," a process by which they slowly lower the freezing point of their bodies.

Because they have so little liquid in them and can't take in more during the winter, drying out is a great danger. That is why many insects spin cocoons, form galls, bury themselves in the ground or under leaf litter, or squeeze behind tree bark. All these techniques slow the evaporation of the insects' meager body liquids.

The insects' many winter survival techniques make it possible for winter walkers to assemble respectable insect lists if they know where to look. Today I decided to collect and study plant galls.

Galls are the abnormal growth of plant tissues caused by the work of insects or bacteria. No two galls are alike. They can be globular, semiglobular, cylindrical, or disklike swellings on leaf blades, leaf petioles, or twigs. More than two thousand kinds of insect galls have been identified in North America alone. The majority of these are made by wasps (805), small, delicate flies called gall midges (700), mites, and aphids.

It's not understood exactly how galls are formed, and we don't know the life histories of many of the gall-makers either. Scientists have taken a particular interest in these questions, however, because galls seem to be related to tumorous growth and may be helpful in understanding cancer in human beings.

Most insect galls are constructed by a mature insect that pierces a part of a plant, usually its stem, to lay an egg. Some scientists believe that at that point a drop of fluid is also deposited, which stimulates the plant tissues around the egg into producing a gall. This gall provides the larva, once it hatches, with food and shelter until it pupates and leaves its temporary refuge as an adult insect.

The majority of insect galls are not harmful to the host plant. Moreover, people have made use of various galls throughout history. In ancient Greece, galls were used as lamp fuel. Gall-nuts produced by the oak species *Quercus infectorius* in Asia Minor, Persia, and Syria were utilized in tanning and dyeing and in the manufacture of ink. Galls are also strong vegetable astringents and have been used in medicine.

Most galls are constructed by a single insect, but often a gall may have several creatures living inside it. This is because some insects use galls made by others. If they don't harm the gall-makers, they are called inquilines. If they kill the gall-makers to use the galls, they are parasites. If the original gall-maker has pupated and left its shelter, the empty shell is often used by other insects as protection during the winter.

Gall-making insects are particularly fond of oak trees; more than eight hundred types of galls have been found on oaks alone. The most noticeable are oak apple galls, of which there are one hundred different types. They are tan, papery, golf ball–sized galls that form on the leaves, leaf stems, and buds of oak trees. In winter they're often found on the ground or hanging on the trees like Christmas balls.

Many other insect galls have names, too, such as the oak bullet gall, the blueberry stem gall, the elliptical goldenrod gall, and the goldenrod bunch gall—all of which are common here.

During my walk, I collected dozens of insect galls, but some had tiny holes in them, indicating that the gall-dwellers had already matured and left their protective homes. Others had larger holes, which meant that bird or small mammal predators had eaten the contents.

I brought them all home and carefully sliced them open with a sharp knife. One blueberry stem gall had spider silk in it, and when I opened it, a minute spider skittered out. I trapped it under my hand lens, noticed that it had only seven legs, and realized that it had used the empty gall husk as a safe wintering home. Blueberry stem galls, shaped like human kidneys, are reddish-brown and are constructed by the gall wasp, *Hermadas nubilipennis.*

The best gall specimen I found was the goldenrod ball gall, a round, shiny, dark brown gall that grows on the stems of goldenrod. It is made by the larva of the spotted-winged fly, *Eurosta solidaginis.* The female lays its egg on the stem in late spring. When the egg hatches, the larva burrows into the stem and makes its gall, spending the winter in larval form.

Once spring comes, the larva constructs a tunnel to the outside layer of the gall, returns to its chamber to pupate, then flies out the tunnel and bursts through the gall as an adult fly.

I sliced through the brittle outside shell and the softer white interior to an oblong chamber in the middle, which contained a fat brown larva, just as the books described.

All the oak apple galls I collected had holes in them that were made by predators, and I found the spongy interiors empty of insects. A goldenrod elliptical gall also had a hole in it but it was smaller. Inside the chamber I found the dried pupal skin of its former occupant, the moth larva *Gnorimoschema gallaesolidaginis*. But that was normal for this insect.

In fall the moth lays its eggs on the leaves and stems of goldenrod. The larva hatches in the spring, crawls toward a new goldenrod shoot, burrows into an end bud, and then crawls down inside the stem where it feeds. This causes the plant to form an elliptically-shaped gall around the larva. It feeds until late July, when it bores an exit hole that it plugs up with silk and plant material before it pupates. Finally, in August or September, it emerges as an adult and leaves the gall via the exit hole.

The most attractive gall I found was a goldenrod bunch gall in the leaf bud of a Canada goldenrod. It had been formed by the midge *Rhopalomyia solidaginis*. The growth of the gall stunts the stem and causes excess leaves to clump at the tip of the winter stalk, which resembles a dried up flower. This year our Canada goldenrods were heavily infested by these midges and now the dried patches of these midge-infested goldenrods that gleam in the sunlight look like enormous bouquets of dried, beige flowers.

February

Nature is full of genius, full of the divinity;
so that not a snow-flake escapes its fashioning hand.
Thoreau, Journal

FEBRUARY 1. This is more like a day in late March or early April—warm, clear, and breezy. Once again the earth has defrosted and muddies my boots.

As I walked along Laurel Ridge Trail, a rustling in the leaves alerted me to a Gapper's red-backed vole gathering dried leaves in its mouth. After a few seconds, the vole scurried into a small hole under the base of a laurel stem.

This beautiful little creature is easily identifiable by a broad, reddish band running from its forehead to its rump, and by its bicolored tail, dark brown above and whitish below. Otherwise, its plump upper body is gray and its belly silvery white to light yellow. Alternate names for it are boreal red-backed vole, southern red-backed vole, and red-backed mouse.

It is a quintessential deep-forest creature, adapted to living underground year-round. There it eats subterranean fungi, a staple in summer and autumn, or ventures above ground in search of seeds, mosses, ferns, berries, lichens, and some insects. The vole's consumption of underground fungi is especially important to the survival of the forest. Most trees de-

pend on the filaments of the fungi to obtain water-borne nutrients from the soil. But the fungi do not produce aboveground fruiting bodies to carry their spores and must rely on
small mammals to disperse them. After being eaten, the
spores of the fungi pass unchanged through the vole's diges-
tive tract, reinoculating habitats where forests have been
logged or destroyed by natural forces.

Dr. Joseph Merritt, the former Resident Director of Powdermill Nature Reserve, which is the biological field station of
Pittsburgh's Carnegie Museum, has been studying the wintering survival of small mammals, including red-backed voles.
After twelve years of live-trapping them, he found that they
range in density from five to thirty-six individuals per hectare.
In autumn and winter, they reduce their body size so they
need less food, shifting their food preferences to readily available seeds, roots, bark, and plant parts, foraging under the
snow and in subterranean burrows where they are not affected by snow or bad weather and where it is warmer. But
because of the unusually warm weather, the one I saw had
ventured aboveground in search of food.

Red-backed voles also engage in nonshivering thermogenesis (heat production) or NST to stay warm. NST is an important winter survival technique, not only for red-backed voles
but for short-tailed shrews as well. Between their shoulder
blades, near their spinal cords, they have high energy, heat-
producing tissue called brown adipose or brown fat, which
functions like a blanket to keep them warm.

Despite all these adaptations, red-backed voles lose weight
throughout the winter. Many die, especially during winters
with little or no snow cover.

In Pennsylvania, those that survive begin breeding in late
March and continue through November. The female builds a
globular nest lined with grass, dead leaves, and sphagnum or
other mosses, which she places in a natural cavity or the abandoned hole or nest of some other small mammal. The nest
accommodates the female, her four or five young and her

male consort. Although he does not help care for his offspring, the male stays with the female through the seventeen-
to nineteen-day gestation period and until the young grow
large enough to fill the nest.

The female weans her young at seventeen days, after starting them on solid foods three days previously. From blind,
toothless, hairless, and pink-skinned at birth to completely
furred and able to climb, run, and groom themselves at two
weeks old, they reach sexual maturity in three months. A pro-
lific breeder, one red-backed vole produces two to three litters each year. The many offspring help to assure the survival
of this species.

This was the first time I ever observed one in the winter,
but I have had two other close encounters with them. One
occurred on the stony top of the Sapsucker Ridge Trail in the
summer. It was nosing under the forest leaf duff and when I
approached, it began circling rapidly from left to right and
jerking its head. It then alternated dancing with searching for
food under the leaf litter and eating. As I counted the number
of spins, the vole gradually increased its revolutions from four
to sixteen. Red-backed voles engage in this "waltzing" behavior, as the scientists call it, whenever they are disturbed.

I again encountered a Gapper's red-backed vole several
summers later, while sitting quietly at the base of an old black
cherry tree. The creature emerged from a small hole in the
ground about eight feet away. It sat up on its haunches and
sniffed the air for a minute before slipping under the leaf duff
to retrieve an acorn. Holding the acorn in its two short front
paws, the vole ate it. Then it looked for and retrieved a second
acorn, which it also ate, finishing off its meal by cleaning its
face with its front paws. Finally it ran along a fallen log, leaped
to another, and disappeared.

Three sightings in thirty-one years makes every one special
and I was pleased to see the Gapper's red-backed vole out on
such a fine winter day.

FEBRUARY 2. Groundhog Day dawned clear and cold, so of course Punxsutawney Phil did see his shadow. Six more weeks of winter. But even when he doesn't see his shadow, we always have at least six more weeks of winter.

By midmorning I was wandering through the spruce grove still searching for the mysterious bird I had twice seen fly from the area. Suddenly, a bird fluttered off. It was about the size of a mourning dove, maybe larger, but its wings did not whistle when it flew. Because I had the feeling it had not left the grove, I circled the trees from below and entered a small opening surrounded by spruces. On one tree branch I spotted a fluff of feathers. I was sure it was an owl, but what species was it? It was too large for a screech-owl and too small for a great horned.

I "pished" softly, and slowly its head swiveled around. At the same time its ears seemed to grow, standing up higher and higher. Its facial disks were a rusty red and between its dark eyes, there was a pattern of black and white that made it look almost catlike. It was a long-eared owl, the first one ever recorded on our property.

I talked quietly to it in what I hoped was a soothing voice, thanking it for being there, for making my day, and for gracing the grove, all the while I peered at it through my binoculars. Then, still talking, I reached down slowly to pick up my walking stick. The owl blinked sleepily. I bid it adieu and rushed back home, absolutely jubilant, and told Bruce and Dave where to find the bird.

Dave set out immediately and located it in the same place. Bruce went up after lunch, lugging his camera and tripod, and the owl held still for him, too, as he took slides to document the 167th bird species here. I was hopeful that we might have a permanent winter resident because long-eared owls like dense conifer groves beside open fields. They especially like to eat deer mice, meadow voles, and shrews, all of which our land supplies in abundance. Other animals they eat that live here are small birds, juvenile rabbits, star-nosed moles, long-tailed weasels, and ruffed grouse.

Because long-eared owls often roost communally in the winter, I wondered if more than one was in the spruces. Roosts typically range from two to twenty birds, although as many as one hundred have been seen. Sometimes, these secretive owls that inhabit dense groves in the winter use the same groves for breeding. They often nest in American crows' nests or close to them, and crows had nested in our spruce grove last spring.

Whether or not it stays it was a wonderful Groundhog Day gift. Who cares if winter lasts another six weeks!

Our eldest son Steve, who studies politics on the side, had a different take on this holiday. He sent us his thoughts by e-mail from Cornell University, where he is working on his Ph.D. in South Asian linguistics. "I was reflecting on the way to school this morning that Groundhog Day is one of my favorite holidays. This is because nobody has yet figured out how to politicize it. There are no Dead White European Groundhogs to criticize; nobody gets exercised about Groundhog displays in public buildings; Groundhog Awareness seems like a lost cause, and there won't be Save the Groundhog or Take Back the Groundhog campaigns any time soon. I suppose at some point it may occur to someone that Groundhog Day is Pagan and Pre-Christian and an insidious influence on our children. But until then, it's a refreshing break from an otherwise unbroken parade of holiday-cum-cause-celebres exploited by politicians and interest groups."

Dave and I thought that Steve's remarks were naive in light of what we had been hearing about the holiday and Dave answered Steve by saying that "there are at least six different 'official' groundhogs looking for their shadows at various locales around the country, with (obviously) conflicting results. Besides our own Punxsutawney Phil, there's Buckeye Chuck

(Ohio), Sun Prairie Jimmy (Wisconsin), Gen. Beau Regard Lee (Georgia), Staten Island Chuck, and even a rival Pennsylvania woodchuck in the Poconos.

"With all this, I'm not sure one can safely maintain that February 2nd offers 'a refreshing break from an otherwise unbroken parade of holiday-cum-cause-celebres exploited by politicians and interest groups.' At least we can be grateful that, thus far, it's only states that are competing for a share of the lucrative groundhog market. It's probably only a matter of time, however, before some multinational corporation decides to take advantage of the cheaper and more compliant marmot labor force overseas."

Needless to say, Dave too studies politics on the side and such lively exchanges are a feature not only of e-mails with our sons but at our own dining table. That's because Bruce, who is spending his retirement researching and writing about peaceful peoples in the world, also keeps up with both national and foreign affairs. By osmosis and living with four such males over the years I too am affected. Winter, after all, is the time to ponder the state of the world and hope that maybe this century will be better than the last one was for all humanity and what is left of the natural world.

FEBRUARY 3. Now that the snow is gone for a time, I set out to look for the most primitive of earth-based plants—lichens and mosses. Lichens consist of two organisms that live together for mutual benefit—a filamentous fungus and a one- or few-celled blue or blue-green alga. The fungus supplies water and minerals that the alga absorbs and turns into carbohydrates. Together they form a new plant body called the thallus.

As pioneers in the harshest environments, lichens range from the highest mountains to rocks near the north and south poles. They also live on tree trunks, decaying wood and soil. Lichens grow slowly; some colonies are thought to be more than two thousand years old.

Although lichens secrete powerful chemicals that can break up rocks, and can live through years of drought because of their gelatinous surface, which forms a protective skin, they will not grow in burnt-over areas or in areas with polluted air. So sensitive are they to pollution that scientists can estimate the amount of sulphur dioxide in the air by studying tree lichens.

Lichens are classified into three types. The most ancient, primitive, small, and inconspicuous ones are called "crustose," meaning "flaky or crusty," and grow flat on rocks or tree trunks.

Probably the most famous crustose lichen is what has been named manna lichen because some scientists speculated that it might have been the Biblical manna from heaven. A desert species, it grows loosely on rocks and is carried by wind and rain. Script or writing lichen, which looks like black scribbles on hardwood tree trunks, is a crustose type that lives on our trees.

"Foliose," meaning "papery or leafy," lichens are ruffle-edged mats of varying colors attached to rocks and trees, such as the rock tripe and shield species. Toad skin lichen, a warty-looking black rock tripe, and pale shield lichen, a leafy-shaped, grayish-green lichen that grows on trees and logs, also live on our mountain.

While most people find it easy to overlook the crustose and foliose lichens, the stalked or branching fruitcose types are easier to admire and identify. Several have been misnamed as moss—reindeer moss is really a lichen, as is British soldiers' moss. The latter is also called red crest lichen because its gray, branched stalks are topped with round, red, fruiting tips. To-day on our powerline right-of-way I searched for and found many showy colonies of British soldiers' lichen as well as the green-branched Iceland lichen and the pale green pyxie cup or goblet lichen. I picked up a couple of rocks, and, through my hand lens, looked at beautifully-colored spots of crustose type lichens starting to break down the rocks.

On the moss-covered trails of Laurel Ridge, mats of shrubby gray reindeer lichen and Iceland lichen thrived. Many of our tree trunks were studded with an array of lacy, gray, foliose types.

Most lichens reproduce either by small fragments breaking off and growing new plants, or by producing dustlike soredia that consist of a few algal cells surrounded by a tangle of fungal hyphae (threads), which are carried by the wind, rain, or feet of animals. When I knelt in a bed of Iceland lichen on the powerline right-of-way to look at it more closely, pieces of it clung aggressively to my wool slacks, and I picked it off with as much effort as I would sticktights, incidentally sowing more Iceland lichen along Laurel Ridge Trail.

Lichen is Greek for "leprous" because the Greek physician and founder of medical botany—Pedanios Dioscorides—back in C.E. 68 thought lichen resembled the skins of lepers and could be used to treat leprosy. Although lichens did not cure that disease, certain lichens in Europe produced an extract that inhibited the growth of tuberculosis bacteria. Other lichens were used to make beer and alcohol, a dye that repelled moths, and as a base for an antibiotic salve. Mites, caterpillars, wood lice, snails, and slugs eat wet lichens of many species, while reindeer lichen is relished by reindeer and used as survival food by humans.

Most important, lichens prepare the substrate on which mosses can grow, by absorbing minerals from rocks and carbon dioxide from the air and collecting dust and particles of crumbled rock until pockets of soil form that enable the spores of mosses to germinate. That is why lichens often grow with mosses even though they are not related.

Unlike multicolored lichens, mosses come only in one color, but in every possible shade, from whitish or bluish green, to golden or yellowish green, to bright olive or dark green. Mosses prefer moist, shady, and cool habitats and are widely distributed in woods and fields, on rocks, trees, and decaying wood and in swamps, streams, and ponds.

In damp weather mosses grow fast. Their stems die below and grow above to add more plant-supporting soil to that produced by lichens. They also use water for reproduction. The sperm swims through a film of moisture over the plant's surface to reach the nearest ovum.

A moss is defined as any small green plant having leafy stems without flowers and growing close enough to form velvety cushions. The easiest way to identify at least some mosses is to examine their mature spore cases or capsules that grow at the end of slender stalks called setae. Moss spore capsules come in all shapes—balls, eggs, urns, horns, and bills are a few examples. Those shapes, as well as how they are attached to the setae or stems (erect, inclined, horizontal, or drooping) help to identify them. In addition, you can examine the cap covering the young capsule called the calyptra, which usually falls off before the spores ripen, along with the operculum, a lid on the upper part of the capsule that falls off when the spores are ripe. The operculum can be convex, cone-shaped, short-beaked, or long-beaked. Underneath it and surrounding the mouth of the capsule is a tiny single or double fringe called a peristome. It has red-to-orange-to-yellow filaments, like teeth, that are sensitive to moisture and tightly closed during wet weather, so spores aren't washed out into the moss from which it has germinated. But when the weather is dry, the teeth of the peristome separate and bend back so the spores can sift out and be spread by the wind. The number of teeth, their size, shape, and character are a way to identify mosses under a microscope.

Moss habitat also helps to identify it. Knothole moss, for instance, grows in knotholes in living hardwood trees. Color can also be a clue. White-tipped moss has leaves with white tips and grows on exposed rocks on ridges. The common pincushion moss forms luxuriant, whitish-green cushions in our woods on both the rocks and soil. Another common and easy-to-identify moss that grows on our trails is one of more than one hundred species of hair cap mosses—*Polytrichum*

ohioense. The capsules are angles and gradually taper to the setae, which are reddish below and yellow above.

Forty to fifty species of mosses can be found in a typical
Pennsylvania woodland and Pennsylvania has over three hundred species in all. But without a microscope and the expertise of a bryologist (one who studies mosses and closely
related liverworts), I've only been able to find a few species
here. Still I must agree with Englishman John Ruskin, who
praised mosses as "the soft green beds into which our feet
sink." I know my feet did today as I walked our lushly-
covered mossy Laurel Ridge Trail and sat at the base of a tree
encircled by pincushion moss.

FEBRUARY 4.   How I miss the snows of past winters.
Today, as I walked along the Far Field Road, I heard the call
of a hairy woodpecker and remembered the cloudless winter
day, several years ago on this date, when I snowshoed along
the Far Field Road. On that day I paused as a male hairy
woodpecker landed on a nearby tree trunk.

The road was sheltered and warm, and the woodpecker,
unlike most of its kind, did not seem to care that I was close
by. This presented me with a perfect opportunity for wood-
pecker-watching and winter sunbathing.

I carefully lay down on the two feet of frozen snow—my
head on my binoculars' case, my back on my ski jacket, my
knees bent, and my feet still attached to my snowshoes. In
this position, I had an excellent view of the foraging woodpecker while I basked in the sun.

He moved up and down the broken-off, dead branch of a
black cherry tree, stripping off large pieces of bark with his
bill. He also used his feet to knock off loose bark. I studied his
long, rapierlike bill, the red spot on the back of the black-and-
white head that distinguishes a male hairy from a female, and
the pale yellow bristlelike feathers that protect his nostrils
from wood dust.

Hairy woodpeckers look like overgrown downy wood-peckers, $9\frac{1}{2}$ inches to the downies' $6\frac{1}{2}$, but hairies have much longer bills and lack the scattering of dark bars on their white outer tail feathers that downies have. Otherwise, both birds have white backs and breasts, black-and-white striped heads, black wings dotted with white, and black inner tail feathers.

They also possess different temperaments. Downies are much more trusting and come readily to backyard feeders filled with suet. Hairies are usually shy, reclusive birds that like large tracts of deciduous forest and rarely venture into back-yards or come to bird feeders. But there are always exceptions to the rule, and I was obviously watching a bolder-than-usual hairy woodpecker.

I watched the bird for about half an hour before he finally lowered his head and pushed off from the tree with his feet, looking like a swimmer kicking off from the edge of a pool.

Today continued to be a woodpecker day. When I returned home, a female pileated woodpecker was sitting on the ground at the base of a dead pine tree in our yard and survey-ing the base of it. First she turned her head toward the ground and then against the base of the tree, as if she was lis-tening for something. Pileateds are known to use the sound of their prey, such as carpenter ants, to track them down in dead wood. She poked at the base for a few minutes, moved on up the tree trunk, sheared off a loose bit of bark, also in pursuit of ants, and then flew off.

In midafternoon, she landed in a dead tree along the woods' edge and began tearing away at the bark. We put the scope on her at the bow window and had an eye-filling view of a pileated at work. We watched her pull three long white grubs from the bark in fifteen minutes. She constantly looked around and stared directly at the bow window as if she knew she was being watched. While I worked at the computer in my study I could glance out and see her moving systemati-cally up and down the tree trunk. Then she was gone.

The pileateds we watch in the surrounding forest hold on to their territory and mates throughout their lives. Many pairs even forage together and often, when I see one pileated, very soon another one appears or I hear it calling in the distance as if maintaining contact with its mate. Because we leave all standing trees in our forest to finish their life and death cycle unmolested, pileateds have plenty of places in which to roost and nest. Their huge feeding excavations in trees also provide food for red-bellied, downy, and hairy woodpeckers and northern flickers.

Their crow size and red crests easily distinguish them from any other woodpecker species. So does their penetrating, percussive drumming and loud "cuk-cuk-cuk" calls. The late George Miksch Sutton, who was Pennsylvania's state ornithologist back in the late 1920s and early 1930s, described their drumming here as "an introductory, rapidly given roll; then a pause, followed by three distinct blows; another pause; and two concluding blows," although they do present other variations on that theme.

They nest in large, dead trees and both parents incubate the four white eggs, but the male does night duty. After about fifteen days the eggs hatch and both parents feed and share brooding duties. Again the male takes the nights. After about twenty-six days the nestlings fledge, staying on with their parents until the fall when they disperse.

The name pileated means "crested," but other, more colorful names include "log-cock," "cock-of-the-woods," "black woodcock," "stump-breaker," "laughing woodpecker" (because of its cackling cry), and, in Juniata County, Pennsylvania, back in the 1920s, "cluck-cock."

Sutton once shot a female at Aitch, Huntingdon County, and found in her stomach 153 carpenter ants, a beetle, a bug, and seventeen wild grapes swallowed whole. The crop and stomach of a male he shot and examined in Northumberland County contained 469 carpenter ants. Other early researchers found as many as 2,600 ants in one stomach and most of those ant species lived in dead trees. Which is why we are never without these marvelous birds and why they are dependent on mature forests.

FEBRUARY 5. A breezy, clear, cold day. Three American crows harassed a mature red-tailed hawk sitting on a tree limb overlooking First Field this morning. As I crossed the driveway for a closer look, the bird flew along the edge of the field and two crows chased and repeatedly dived at it. The scene reminded me of a bee stinging a black bear. Again the red-tail landed on a tree limb; again the crows flew in close to voice their displeasure; and again, when I started walking up the field, the red-tail took off, this time up over Sapsucker Ridge where it hovered on outspread wings, then dove down behind the ridge, closely followed by a second red-tail that seemed to have appeared from nowhere. The crows, having done their duty, flew across First Field.

Otherwise the woods were silent except for two ruffed grouse that flushed below the First Field Road and the far-off calls of black-capped chickadees. But if I look hard, I see other animals and birds in stumps and branches, boles and leaf nests, fallen trees, vines—wherever something looks a little different in the landscape. As winter continues open and empty, I picture the landscape more and more with imaginative figures conjured up from oddly-shaped, inanimate, natural objects. Winter is a time when "young men dream dreams" and so, in my case, does an aging woman still young at heart.

FEBRUARY 6. I went outside before dawn to retrieve the feeder containers from the back porch. A striped skunk, its tail streaming out straight behind, as if it wouldn't think of threatening anyone, disappeared under our front porch. I caught not a whiff of skunk odor. But when I returned from the basement with birdseed, I put on the porch light and carefully looked out. The skunk was back, eating seeds on the sidewalk in front of the bulkhead door, its white muff back and tail gusting in the breeze. It ate as if starved and only retreated when the first birds appeared in the dim light of predawn. Although skunks do sleep in their dens, living off their accumulated fat during severe winter weather, they emerge during mild spells in search of food.

A sharpie sat on the ground outside the bow window when I came downstairs after doing my exercises. It remained still as if in contemplation for at least fifteen minutes before starting to pluck the dead, dark-eyed junco it had pinned to the ground. Eventually the raptor moved behind the juniper tree, still sitting on the ground, and by 9:00 A.M. black-capped chickadees and tufted titmice ignored it and flew in and out of the feeder for seeds. No ground feeders, including juncos, returned, however. Finally, I stepped out on the back porch, figuring that would scare the sharpie off. But even as I talked to it, it remained seated. I dressed to go outside, went back out on the back porch, and walked toward the raptor. At a distance of fifteen feet, it flushed, and off it flew over the yard, accompanied by a steady buzz of scolding from all the little birds in the trees.

As I walked across First Field, I heard the steady calling of American robins from the sunny slopes of Sapsucker Ridge, which continued along Bird Count Trail. No wonder they liked this place. I sat on the trail and was protected from the cold wind and warmed by the sun. A snow shower had dusted the mountain white so it again looked like winter.

As I continued on Bird Count Trail, I found the mother lode of robins on grapevines or flying overhead—hundreds of them and a couple were even singing! I also discovered a white-throated sparrow and American tree sparrows on blackberry canes. A male hairy woodpecker flew in to forage on the trail. Carolina wrens sang in the distance. A blue jay called as a pair of red-tails sailed past. Cedar waxwings perched in a privet shrub eating berries.

Robins continued flying up and foraging in the ground above and below me as I walked the Greenbrier Trail. Hundreds more robins foraged in the Bench Blind Hollow along with four ruffed grouse, a pileated, and a red-bellied woodpecker. Bruce reported many robins along the Far Field Road too.

He or I or Dave have checked the spruce grove every day in search of the long-eared owl, but it is gone. When I took a quick walk up there after sunset, I startled several deer along the way. Then I sat on Alan's Bench and listened to a chorus of high notes from dark-eyed juncos coming in to roost for the night. Since I heard them calling long before I reached the grove, I don't believe they were alarm notes at my arrival but greetings to their fellow spruce roosters. I also heard the lower "ticks" of northern cardinals and startled at least two silent American crows from the trees. It is truly a beautiful time of day and as I walked down through First Field in the gathering dusk, a couple of song sparrows flew up briefly from the base of the dried grasses, where they must spend the night.

FEBRUARY 7. While New York and New England have record snow packs and even eastern Pennsylvania had as much as fifteen inches of snow the other day, we continue to get little or nothing.

On this day it was a breezy twenty-two degrees Fahrenheit and patches of bare earth alternated with patches of frozen snow. I stopped to watch a northern short-tailed shrew foraging on the edge of our powerline right-of-way. The shrew had scuttled past a mere five feet away. Then it paused and used its long, mobile, cartilaginous snout to poke in leaf litter and dried grasses in search of food. Next it pushed its snout under a snowy patch for ten minutes and busily ate whatever it had found.

That was when I slowly eased myself down on my "hot seat" to watch it. The shrew was too close to focus my binoculars on and remained oblivious to my presence. It pursued its prey vigorously, its pointed snout questing, its clawed back feet pumping, its front feet digging like a frantic terrier. Once it pulled what looked like a caterpillar from beneath the leaf litter and chomped it down.

A small, plush, charcoal-gray, furry ball, it scuffled over the snow. Its pink nose constantly sniffed while its naked, pink feet scratched the thin snow layer or the open turf. The little creature ate so much that it even paused to excrete.

After almost forty-five minutes of high-octane hunting and eating, the shrew ran under a log at the edge of the woods. Probably it was returning to its apple-sized resting nest. Constructed of grasses, sedges, and leaves in the shape of a hollow ball, the nest is located as much as six to sixteen inches below ground or beneath logs, stumps, or old boards. From the nest, openings lead to a complex underground burrow system that includes separate food-caching locations and latrine areas.

Mostly northern short-tailed shrews sleep in the winter to reduce their need for food. But such periods are alternated with intense active hunting periods that usually occur below the snow cover where it is warmer. Researchers claim that northern short-tailed shrews spend only brief periods above ground during cold weather, but this one, at least, was undeterred by the cold.

I walked on to Coyote Bench where I watched the slow ballet of squirrels up and down trees and branches, the male trailing the female, sniffing her rear end and the branches she moved over to see if she was ready to mate. But eventually she wedged her rear firmly against the trunk of a large snag and sat on a branch for long, quiet minutes, grooming herself. She had a large splotch of brown just above her white belly halfway along her gray sides, which made her easy to identify. Next she climbed quietly up and down several trees and finally disappeared.

FEBRUARY 8. Twenty-eight degrees and lightly snowing at dawn. All the little birds were eager for food so I filled the feeders before making breakfast. At the feeder this snowy day we had a male northern cardinal, a red-bellied woodpecker, and many American goldfinches. Late in the afternoon, when I went downstairs to brew tea, I found the sharpie sitting on the porch below the feeders while a male white-breasted nuthatch was frozen in an upside-down position on the inside wall of the wooden feeder. The sharpie flew when it saw me. A few minutes later Dave spotted it sitting on the ash tree overlooking the back porch. Again it flew when we saw it. All the while the nuthatch remained frozen in place a good ten minutes and no other birds came near.

"The white-breasted nuthatch," Winsor Marrett Tyler once wrote, "is a droll, earnest little bird, rather sedate and unemotional. He is no great musician and seems to lack a sense of humor. . . . he appears to take life on a matter-of-fact level."

For those of us who do most of our observing of white-breasted nuthatches at our bird feeders or in a winter woods, Tyler's description is on target. Even when the birds walk down tree trunks in search of food they manage to maintain their businesslike demeanor and they hold themselves aloof from other birds both at the feeder and in the woods.

Frequently they join mixed foraging flocks of tufted titmice, downy woodpeckers, and black-capped chickadees. But often the nuthatches chase the other birds away from food items. One December, for instance, I watched a nuthatch rout a downy woodpecker from its tree branch. They also threaten other birds at the feeder, including other nuthatches, by spreading their tails and swaying back and forth, or they clear the feeder of lesser birds with thrusts of their rapierlike bills.

Their plumage is sober and conservative: white breasts, gray backs, black heads—winter colors for winter birds. Their songs, such as they are, are equally dull—nasal, low, and all on one pitch.

Because they are so common, I tend to overlook them. Unlike many bird species, nuthatches are increasing in numbers. That's because they are not particularly choosy about their habitat requirements, living in much of the northern hemisphere from southern Canada to southern Mexico.

They also have catholic food tastes, eating a wide variety of nuts, seeds, and insects. Access to bird feeders, studies have shown, improve their nutritional condition. And their technique of scatterhoarding food during the winter helps to ensure a steady food supply. They use each storage place only once, stuffing single food items in bark crevices on the trunks of large trees and on the underside of branches. Sometimes they cover the food with a piece of bark or rotten wood, lichens, moss, or even snow.

They prefer trees with deeply furrowed bark, such as chestnut oak, for caching. Although females are not choosy about their caching sites, males prefer to store food on tree trunks. Sunflower seeds are popular items for caching, particularly shelled ones. They also like beechnuts and cache them intensively in fall when they are available. As bird feeder owners can easily observe, most caching is done early in the day and then decreases later.

The sex life of white-breasted nuthatches is as conservative as their plumage. They mate for life, stay together throughout the year and don't indulge in "extramarital" affairs, as far as researchers can determine. The male dutifully renews his pair bonds every year, beginning in January, by staying within fifty feet of his mate, singing from the treetops in early morning, and performing his song-bow by extending his neck and bobbing his head and body every time he sings. Often he sways from side to side between songs.

Most of the year the male dominates his mate, but in late winter, instead of threatening her if she dares to land on the feeder when he is there, he begins flying to her and feeding her. Mate-feeding, as it is called, increases throughout spring and continues until after the young hatch.

As feeder watchers, it is sometimes possible to see these interactions on the feeders or on the trees in our yard from now until late spring.

Today the nuthatch finally broke his frozen pose and went back to eating when the first chickadee flew into the feeder, providing the all-clear signal he needed that the sharpie was truly gone—at least for now.

FEBRUARY 9. Most of yesterday's snow melted in the warmth of the afternoon sun. It was fifty degrees, so I changed into spring clothes and jacket before heading outside.

This was the kind of day when insects might be out. The easiest ones to see were the winter crane flies or snow flies, which creep upward from hiding places at the bases of tree trunks as the temperature rises during snowstorms. Once the sun shines, the light entices them to crawl over the snow in search of mates. After copulation, the females bury beneath the snow to deposit their eggs close to tree trunks.

Snow fleas are another story. Because they are miniscule, wingless insects that grow no bigger than a tenth of an inch, they are harder to see. They overwinter as adults and are active on the surface of the snow or on the top of leaves, but I have to look carefully. When I see small flecks of black appearing and disappearing on the snow, I know I am looking at snow fleas.

Although snow fleas spend bitter winter days under the leaf litter, feeding on decaying vegetative matter, on mild days they may move eighty feet, over a couple of days, in groups as large as half a million to a million. No one knows why they do this. After a few days the groups disappear and scientists speculate that this is a dispersal mechanism. Other snow fleas, encouraged by the warmth, merely jump about and then return at night to the same place where they emerged in the morning.

They mate in late winter and spring, and the females lay

their eggs beneath the soil surface. Those eggs hatch and the young feed throughout the summer. By winter they are full-fledged adults.

Then there are the winter moths, fifty species of owlet moths in North America that are active throughout the winter. We often see them flying when we drive up the hollow road at night. Unlike most insects, they do not produce antifreeze. Their blood freezes at between thirty and twenty-eight degrees Fahrenheit, just like summer insects. For this reason, they venture out only when the temperature is thirty-two degrees or higher. Otherwise, they spend their time beneath fallen leaves where even at night—and on snow-covered ground—the temperature rarely falls below thirty-seven degrees.

All of these winter-hardy moths are endothermic; they can elevate their body temperature by their own metabolism, usually through shivering. In addition, these moths have thoraxes (where their flight muscles are located) covered with dense hair. They also possess a unique circulatory system. As their blood flows out of their thoraxes, it gives up its heat to the blood flowing back in. This produces flying temperatures similar to those of summer moths, between eighty-six and ninety-five degrees.

These winter moths feed exclusively on sweet tree sap oozing from sugar maples and birches. Scientist Bernd Heinrich, who has studied these moths extensively, lured them into his research site by mixing one can of beer, three-quarters of a pound of sugar, some molasses, and a little mashed fruit and smearing it on tree trunks. He put the moths he caught into vials of water, froze them into a block of ice in his freezer, then took them out and let them thaw.

"Once released from the grip of the ice," he wrote in a February 1994 *Natural History* article, "the moths righted themselves, shivered for a few minutes, and then flew off."

But why are they adapted to thrive in the winter? Heinrich

wondered. A matter of survival, he discovered. They lay their eggs before the tree buds open and the larvae hatch and start feeding the moment new leaves appear. By the time predatory migratory birds return, they have completed both their larval and pupal stages. Then they drop to the ground, bury themselves, and estivate for the summer, safe from bird predation.

Now, whenever I see a moth flying on a winter night, I no longer think I am seeing things. Some insects do like it cold.

FEBRUARY 10. Another day that is perfect for walking down our hollow road. The eastern hemlock tree branches are bowed down with snow and the dried beige leaves of the American beech trees add color to the scene. Whenever I walk in our forest, especially in the hollow with its large and diverse tree species, I think about the dream Bruce and I have. We want our 640 acres of mountain land to not just recover from nearly two centuries of misuse, but to develop into an old-growth forest. We are, in short, tree-huggers. We visit patches of Pennsylvania old-growth forest to gain serenity and to slip back into an earlier time, "when deep ravines, where pines, beeches, chestnuts, birches, maples, and walnut trees of various kinds form[ed] a gloomy forest, and fallen and decayed trunks check[ed] your advance at every step; cool, sylvan brooks rushed foaming through all the defiles, and we had to continually cross them on natural bridges, formed by the fallen trunks of trees. Such old trunks are covered with a world of mosses, lichens, fungi-wood, sorrel, ferns, etc., nay, even younger shoots of maple, beeches, and tulip trees, had taken root on them." Prince Maximilian of Wied wrote those words back in 1832 when describing the uncut forests of nearby Cambria County, one of the few such descriptions of Pennsylvania's old-growth forests before the woodcutters moved through.

Unlike the old-growth forests of the Northwest, a portion of which has not been cut and so can serve as a model to scientists for what old growth in that area should contain, only a few remnant patches provide us with clues to what eastern old-growth looked like or what life forms disappeared with the forests. Like the rest of the cutover eastern forest, even those remnant patches lost wolves, elk, mountain lions, martens, fishers, and probably countless small creatures, shrubs, and wildflower species.

Since those days when Pennsylvania was clear cut, a new forest has grown up, one with more hardwoods and less conifers than the original. At least one tree species—the American chestnut—has been nearly extirpated by the chestnut blight imported from Asia and our oaks have been reduced by gypsy moth caterpillars, another foreign invader.

New foreign insects and diseases, as well as native ones, continue to plague our forests, and some researchers are finding that acid rain is harming forest soils. Too many deer eat the understory in some places; hay-scented ferns crowd out new tree seedlings once an area has been cut. With so many problems besetting our forests, our dream of an old-growth forest may be naive.

For instance, our dogwood trees are threatened by a fast-moving blight called dogwood anthracnose, a leaf fungus that first appeared in the Pacific Northwest in the native Pacific dogwood, in 1979. Simultaneously, it was reported in our flowering dogwood, in southeastern New York and southwestern Connecticut. Although a similar species, in the same genus, had been a weak parasite on dogwood, the new strain moves quickly, attacking both seedlings and adult trees in the cool, moist understory of eastern forests where wild flowering dogwood lives. Those in sunny places such as gardens and parks live longer.

Once scientists realized they were dealing with a new species, they gave it a new name—*Discula cestructiva Redlin*—for Scott C. Redlin, who, after an exhaustive search through the scientific literature, where he found no mention of the leaf fungus, and many hours of observing it under an electron microscope, determined that it was a new species.

Ironically, our best stand of flowering dogwoods survived the clear-cutting on what we call Dogwood Knoll. By removing the more than one hundred-year-old red oaks that shaded the knoll, the logger may have inadvertently prolonged the life of our dogwoods. But not forever. Unless a miracle occurs, our dogwood-graced springs are numbered.

But the hollow area is a special place, the heart, as we see it, of the old-growth forest we hope to grow here. Already aging trees shelter birds and mammals from the storms and provide food for them. Deer file down to the stream for water and ruffed grouse perch in the hemlock branches. During a white-winged crossbill invasion one winter, the hemlocks were filled with the beautiful birds eating seeds from the hemlock cones.

But alien invaders may soon be stalking our beeches and hemlocks. The hemlock woolly adelgid *(Adelges tsugae)*, an Asian import, is a pinhead-size, reddish-brown insect that attacks during the winter. Able to survive extreme cold, it sucks the hemlock's sap while simultaneously injecting a toxic saliva. As a result, the hemlock's needles and branches dry out and discolor and the trees die in one to four years. One scientist discovered that woolly adelgids seem to increase when more nitrogen is present in the trees and wondered if added nitrogen from air pollution might be part of the reason for the insect's rapid increase.

Then there is the beech bark scale insect *(Cryptococcus fagi)* that came into the United States from Europe in the late nineteenth century but only became a problem after World War II. In two to five years, the sucking insect pits the thin bark of the beech trees, allowing a deadly fungus to enter the cambium layer and the sapwood and quickly kill the trees. Like the American chestnut tree, new beeches spring up from the roots of the dead trees but are killed off when the disease returns.

I can only conclude that our forests are in trouble. Even with all the insect threats, though, allowing our forest to mature so scientists in the future will have old-growth forests to study is still a good idea. But the old-growth forest we dream of needs several hundred years to reach its potential, something we will not live to see. Perhaps, though, a great-great-great-grandchild will.

Another component of our older, unlogged forest is the amount of dead wood both in standing snags and fallen timber. To many people, our property looks messy, but nature thrives on mess. In North American forests, 120 species of birds roost, nest, or forage on dead wood. So do 140 mammals, 270 species of reptiles and amphibians, and uncounted numbers of *native* insects.

"Dead wood contributes to the vigor of forests by unlocking the nutrients stored in plant material and recycling them to the myriad organisms essential to forest food chains. Downed logs create organic matter in which seedlings take root. They slow erosion on hillsides by forming a barrier to soils creeping down slope," Jennifer Ackerman writes in the *Nature Conservancy* magazine.

We may be able to do nothing about the insects and diseases attacking our forest or the acid rain impacting the soil. But by allowing all the dead wood to remain in the forest, we may help the forest to retain some resiliency. And as our forest grows older and more diverse, the threat of disease and insects may lessen. For instance, in the case of beech bark disease, most scientists believe that if the beech trees are interspersed with other tree species, the damage from the disease will be limited because the scale has greater difficulty in finding host trees.

Growing an old-growth forest may be the best thing we can do, not only to promote biodiversity, but to ameliorate the effects of insects, disease, and our industrial civilization.

FEBRUARY 11. I was awakened at 4:00 A.M. by howling winds that shook the bedroom door handle. Another skim of snow covered the ground, it was twenty-four degrees and still terrifically windy when I left for a walk at 8:15. But a downy woodpecker drummed despite the cold and wind.

It was a day of snow diamonds sifting down in the sunlight. At the edge of First Field, I flushed a ruffed grouse that was eating wild grapes. I sat in the woods, my back against a cherry tree, and watched as a male red-bellied woodpecker landed on a snag in front of me and tapped away at a hole formed long ago from a broken-off branch, extracted an acorn, and carried it in his bill to a deeply-furrowed chestnut oak tree. He jammed the acorn into a crevice and hammered at it with his long, stilettolike bill. Finally a section of the shell fell to the ground. Then the red-bellied extracted some food for himself before jamming it back into a crevice and hitching farther up the tree where I lost sight of him. To me it appeared as if the red-bellied had extracted an acorn from a previous storage area and carried it to another tree to eat, using the chestnut oak's excellent bark as a vice-grip.

I sought refuge from the wind along Greenbrier Trail and continued on past the Bench Blind Hollow into the forest of Hercules'-club trees. Under them, every striped maple that had sprouted had been nipped off by deer so that the understory was nothing but a crowd of dead, upright twigs, many of which were several feet tall. Black-capped chickadees ate tulip tree seeds near the end of Ten Springs Trail, along with a troupe of tufted titmice.

I went down finally into the hollow road forest where chipmunks were courting near Waterthrush Bench, chasing both above and below the road. I counted at least eight and maybe more. Wind whistled in the hollow, but it was not too cold for courting chippies. They also occasionally emitted a whining call that was new to me. More single chipmunks were out as I headed up the road. They usually do court and mate in February if the winter is mild, but this is the earliest I have ever observed such behavior.

After supper Bruce asked me to walk outside for a few minutes to look at the brilliance of stars and planets in a dark, cloudless night sky. Orion was so clear I could see the entire "belt." Off to the side the clustered stars of the Pleiades glittered. The winter triangle superbly displayed its three first magnitude stars and both Jupiter and Saturn shone as brightly as those stars in the vicinity of Taurus the Bull. I easily traced the pattern of the Big Dipper and the W-shape of Cassiopea in the black sky. One of winter's greatest gifts to stargazers is the clarity of the sky after heavy winds.

FEBRUARY 12. As I sat in my study at 8:45 A.M., I suddenly glimpsed a wild turkey down in the driveway. I rushed to the window and had a marvelous view of six foraging turkeys. Two were young jakes with scrubby beards and one chased another from the area he had already scratched up. Each seemed to forage on its own and they kept moving as they fed from weed heads. Two leaped over the fallen apple tree: the other four went around it. Then they all headed up the slope below the feeders. I rushed to Bruce's study window and watched them mosey behind the old rabbit hutch and through the former garden site. Several minutes later they emerged from behind the garage, and a couple stopped to forage in the fenced garden remnants while the others walked on. One of the laggers-behind flapped its wings and flew awkwardly above the ground for about twenty feet. Then it hit the ground running to catch up, reminding me of a group of older children leaving a younger one behind. Remaining a fair distance from each other, they headed toward Sapsucker Ridge, turned northeast about three-quarters of the way up First Field and continued to move slowly across the cut field area as an ice fog rolled in.

FEBRUARY 13. While on the phone this morning, I glanced out the window and saw a fox squirrel actively gathering and eating black walnuts. Occasionally it climbed up a black walnut tree to sit in a crotch of limbs and eat, but mostly it remained at the base of a tree. It always started nervously when a gray squirrel scampered by and ran up between the forsythia bushes and garage whenever something spooked it. The fox squirrel was around for over an hour before I lost sight of it. Fox squirrels are expanding their range on the mountain, but they always seem to be dominated by the more numerous gray squirrels despite their superior size.

I didn't get outside until after ten in the morning when it had already started to cloud over, but along the Far Field Road I thought I heard, but could not see, Canada geese flying over. Probably they were just the locals and didn't signal the beginning of spring migration.

At the Far Field Thicket, as I passed the old red fox den, I sniffed at the skunky odor emanating from the soil—the smell of mated foxes—so I imagine the den is occupied. The beginnings of a stream oozed from the ground and flowed through the thicket toward the valley. I sat between the stream and the mossy, shrubby miniature wetland below the thicket, enjoying the peace, weak sunshine, and "peter-peters" of a distant tufted titmouse.

The reddish-glow of a moss bed growing on a fallen log caught my attention. The glow was caused by hundreds of spore cases of fern moss, and when I rubbed my hand over the top, greenish dust (spores) erupted in a cloud, turning my hand green. The orange-brown spore cases sat atop thin, deep red stems, which gave an overall reddish cast to the mossy log. Large puff balls on the slope above the Far Field also oozed green dust.

In the Far Field woods a pair of vocal white-breasted nuthatches "duetted"—a long, drawn-out, tremulo version of their usual "yanking" that continued without cessation for several minutes.

Then, when I reached Coyote Bench, the forest came alive with music and color. A flock of cedar waxwings, whistling while they worked, harvested wild grapes from vines directly above my head.

Mature cedar waxwings almost always look like perfect ladies and gentlemen in their sleek, unruffled, reddish-brown coats, accented by their black masks and crested heads. Not a feather is out of place. Their gray tails banded in yellow, wax-like red tips on their secondary wing feathers, and golden bellies add to their overall handsome appearance.

But when they are drunk these dignified birds become disheveled. According to a Massachusetts witness back in the nineteenth century, cedar waxwings eating fermented black cherry fruits "looked like their feathers were brushed the wrong way . . . some tumbled to the ground with outspread wings and attempted to run away. Still others tottered on the branches with wings continuously flapping, as though for balance." They also "kept up a continual hissing noise, such as snakes might do."

Many more accounts of "drunk" cedar waxwings and other birds have been recorded by casual observers, but late in the twentieth century the toxicologists weighed in when several cedar waxwings fell from a rooftop after eating over-wintered hawthorne fruits in central Indiana and died. Although the birds were technically drunk, the toxicologists reported, they died from their fall, not from alcohol poisoning. Unlike humans, such mishaps are due to their preference for fruit, not alcohol.

Fruit makes up 84 percent of the cedar waxwings' diet. Because they are the primary consumers and dispersers of cedar or juniper berries, they were named cedar waxwings. They have also been nicknamed "cedar-birds" as well as "cherry-birds" in recognition of their fondness for wild black cherries. As soon as the cherries ripen here in mid-August, I hear their high-pitched, whistling calls as they feed, hidden in the lush canopy above me.

Their calls are an important device for keeping the flock together, especially when they suddenly take off on long flights
to another fruit source. Their preference for fruit, in turn, dic-
tates their flocking, nomadic lifestyle. Throughout the year
they move from fruit source to fruit source and delay breeding until summer, when seasonal fruits are abundant. They
have no concept of defending a territory. When they breed,
they nest in loose clusters and continue to congregate in
flocks at fruit crops away from their nesting areas.

But each couple is monogamous from spring until late
summer and develops strong pair bonds by performing what
ornithologists call the "courtship dance" or "courtship-
hopping." Usually the male approaches the perched female
with a small item such as a morsel of fruit or an insect in his
bill and hops sideways to her as he gives her his "gift." She, in
turn, hops away and then back and returns his "gift." Again
he hops away, often bowing, and hops back. This "gift-
giving" back and forth is repeated as often as twelve times.
After the proverbial "billing and cooing," they copulate.

Cedar waxwings judge each other's age and suitability by
the number and size of the red tips on their wings. They
don't engage in May-and-December relationships—older females mate with older males and immature males with immature females. After a prolonged courtship (in bird terms),
nest-selection and then building begins in early to late June.
Although the female seems to choose the nest site, usually at
the edge of a wooded area and preferably in a fruit orchard or
young pine plantation near other waxwing nests, they first fly
from potential site to potential site. At each one, she perches
in the fork of the tree or shrub and makes "nest-shaping"
motions.

Both gather the nearest available nest materials, but she
builds the open cup nest in trees and shrubs as low as three
feet and as high as fifty feet from the ground. Then she lays
four to five pale blue or blue-gray eggs splotched with black
and gray and incubates them for twelve days. During that

time, the male feeds her and guards her and her nest from a nearby high perch.

Once the eggs hatch, she broods the naked young while he brings mostly insect food to them for the first three days. Then he adds fruit to their diet. By the time they reach fledging age at fifteen days, they are on an almost wholly fruit diet. If they raise a second brood, the male does the fledgling feeding for a week or more. But the youngsters also form into small flocks with neighboring youngsters as soon as four days after fledging.

Cedar waxwings are the most congenial of birds and I am happy that at least a couple of flocks usually spend the fall and winter here, seemingly undeterred by snow and cold.

FEBRUARY 14. Valentine's Day. I spotted a woodchuck on the barn bank that scooted on up into the weeds as if he were in a hurry. No doubt it was a male, since male woodchucks leave hibernation a full month ahead of females. During this period there is much competition with other males and many go in search of mates, seeking out the females still sleeping in their dens. All woodchucks hibernate here from early November to at least the middle of February. After that, we may see males abroad even when there are several inches of snow on the ground.

During hibernation a woodchuck's heartbeats are reduced from one hundred to a mere fifteen beats a minute. In addition, a woodchuck loses 30 to 40 percent of the weight it gained by eating heavily in early autumn. Once the males emerge from hibernation for good, they still live off their accumulated fat for a month or so, since their minds are more on sex than food. The females need their fat accumulation to help them through pregnancy and nursing their young.

Today, though, is not a good day for anybody to be abroad. The world is encased in ice at thirty-two degrees. Sitting on the veranda, I could hear the continual breaking and falling of branches and trees. Once again our juniper tree was bent over just as it had recovered from the last bad ice storm. Icicles four to six inches long hung from every tree branch, and weeds were sheathed in ice, giving them a rosy hue in the diffused light. The ice had turned the world magical, but deadly, with glitter. Perhaps that is Nature's message—that in death there can be a strange, frightening beauty as the diseased, the weak, and the surplus trees are mercilessly pruned.

By late afternoon, a warm wind had swept Laurel Ridge clear of ice while the sheltered side of Sapsucker Ridge still glowed silver and white. The trees and branches had at last stopped splitting. Occasional bursts of sunlight from behind masses of white clouds momentarily illuminated random trees as I set out to circle First Field. I continued on to Coyote Bench beneath ice-bowed trees and branches, listening to the steady "drip, drip" of melting ice. Many branches and a few trees were down along the Far Field Road. I crunched through icy snow, which heralded my approach well in advance. A hairy woodpecker called, the only creature I heard besides the flushing of a ruffed grouse. As the sun set, the sky turned a soft powder-pink and blue.

FEBRUARY 15. Forty-six degrees by midmorning and sunny. Then a wind picked up, drowning out the noise from the interstate that had been coming right through the walls of the house, so once again, I blessed the wind. It fools me into thinking I live in a miniwilderness where seven common ravens dive and dip in the wind, wild turkeys forage, and eastern bluebirds call.

Bruce washed the road salt and mud off the car while I rushed out in early afternoon to walk on the thawed earth.

I just finished reading *The Poisonwood Bible* by Barbara Kingsolver—a novel about our government's murderous meddling in the Congo and Angola, and I wonder, are humans evil or merely stupid? Or are evil and stupidity the same thing? Why has humanity been allowed by God or Nature to arise and permanently deface creation? Are we, indeed, king-pins in a world of "lesser" creatures? Are there other such kingpins on other planets? If so, are they equally stupid and evil, self-serving and self-absorbed, with here and there a truly good person who redeems the entire species? This may not be the message that Kingsolver wants us to take from her book, but I find it increasingly difficult to reconcile human ugliness with the incredible beauty of the earth that gave us life.

The trees have shed all their ice, and on First Field and the Far Field Road trails, a litter of branches cluttered the ground. The edge of Sapsucker Ridge along First Field contained a line of freshly-created snags, branch after limb after trunk snapped off. Mostly they are black locust and black cherry trees. The First Field Trail contained black birch sapling branches that had broken off, while the major casualties along the Far Field Road were red maple trees.

FEBRUARY 16. Once again I'm counting birds for sci-ence. When I first heard about the Great Backyard Bird Count (GBBC), I was enthusiastic. Instead of only one day, like the Christmas Bird Count and National Migratory Bird Day, I had four days. And it takes place during the psycholog-ically-longest winter month, even though numerically it's the shortest.

Today is the first day of the count, and, much to my disap-pointment, it is rainy and foggy, so I did all my counting from the house. At the bird feeders I counted two black-capped chickadees, a mourning dove, white-breasted nuthatch, an American tree sparrow, a pair of red-bellied woodpeckers, two tufted titmice, two singing song sparrows, and three dark-eyed juncos. In the yard an American crow foraged and a Car-olina wren sang. Even though I may have been discouraged by the gloomy weather, the birds seemed energized.

FEBRUARY 17. The second day of the GBBC is cold, windy, and overcast with heavy snow flurries, but I go out anyway, determined to count birds on our recovering clear-cut. For my trouble, I glimpsed two ruffed grouse, heard an explosive flight of wild turkeys—and found the fresh tracks of three of them in the snow—and listened to a singing Carolina wren undeterred by bitter cold, wind, and snow. One doughty black-capped chickadee was also abroad.

I hadn't dressed properly for the cold, and I forgot my hot seat that I always attach to my belt loop so I can sit down no matter what is beneath me. On top of that, I was suddenly struck by a piercing pain in my back. Usually I sit or lie down when I get such a pain and it goes away, but because I forgot my seat, lying against the snowy bank only aggravated it. Slowly, I pulled myself up and walked more than a mile, mostly uphill, along the steep Ten Springs Trail, snow falling, wind howling, my left hand pressed against the small of my back, stopping frequently to stand and rest. If there were any birds, I didn't notice them.

It took me an hour to get home. By then my sweat pants were soaked and frozen where I had rested against the bank. After an hour of rest, my back recovered, but I did the rest of my bird counting out the window—two song sparrows, eight dark-eyed juncos, two house finches, one red-bellied wood-pecker, three mourning doves, two tufted titmice, two American tree sparrows, and one white-breasted nuthatch. Another ho-hum list. But Dave, who walked down the hollow road to the post office in town, reported a winter wren and a flock of golden-crowned kinglets.

FEBRUARY 18. At last a beautiful day—sixteen degrees and clear at dawn. When I finally got out, at 10 A.M., I headed across First Field and up the warm, south-facing slope of Sapsucker Ridge along Big Tree Trail. Several black-capped chickadees and tufted titmice foraged in the treetops. I heard the quiet tapping of a downy woodpecker and watched as it flew off. Tufted titmice "peter-petered" back and forth, and then a Carolina wren joined in as if competing with the titmice. I sat and shared time with the wren as it perched for half an hour on a tree limb near the ground, and we basked together in a patch of sunlight.

Finally, reluctantly, I continued my walk, hearing nothing but wind as I followed the ridge top to the Norway spruce grove, and then on to the Far Field. It was as if all the birds had fled.

Then, in the woods beyond the Far Field, I heard and saw a hairy woodpecker, a white-breasted nuthatch, chickadees, and titmice. Coming back along the Far Field Road, I detected the high-pitched calling of a golden-crowned kinglet, but I couldn't call it down by pishing. Why is it whenever I count birds for science, no matter how many days that count may be, I never find the numbers and species I usually hear or see on a daily basis, such as common ravens and red-tailed hawks?

In midafternoon, I went out again, beguiled by the sunlight and long shadows, hoping to catch some new birds along Greenbrier Trail. Two crows cawed overhead and I heard a ruffed grouse, assorted chickadees, and titmice. Best of all was a foraging pileated woodpecker.

I sat in silence beneath my favorite double white oak tree on Dogwood Knoll, soaking up the rapidly retreating sunlight. Only the rattling of dried leaves still clinging to the tree broke the winter silence. When I dropped down to our hollow road along Pit Mound Trail, I barely heard the calls of kinglets, chickadees, and titmice over the rushing water of the ice-rimmed stream. Still, I have added two new species to my list today and the feeder juncos increased to twenty.

FEBRUARY 19.   Seventeen degrees and a red sunrise followed by clouds. On this last day of the Great Backyard Bird Count, three Carolina wrens caroled when I stepped outside, one from the side of Sapsucker Ridge, one from behind the shed, and a third from below the guesthouse.

A pair of downy woodpeckers, several tufted titmice, and a ruffed grouse foraged on Bird Count Trail. Two resident Canada geese honked along the ridge top. A male cardinal "ticked" on and on in the greenbrier and then flew down across the trail in answer to another tick. Fifteen juncos twittered in a nearby tree and I glimpsed the flashing black-and-white wings of a pileated woodpecker as it flew overhead.

Below the bench blind more Carolina wrens sang and two female cardinals foraged. A golden-crowned kinglet accompanied a couple of black-capped chickadees and obligingly responded to my pishing by hovering close to my face.

Later, a blue jay greeted me in a black walnut tree in the yard and a pileated woodpecker probed in the old apple tree behind the guesthouse. It was, primarily, a pileated day, because a pair also flew into the yard after lunch. The magnificent, crested male drummed on an old grape arbor fence post, and I watched him through binoculars from our bow window.

At dusk Dave heard the quavering cry of an eastern screech-owl outside the guesthouse. Altogether, we have counted twenty-two species, the same as last year, despite my spending more hours outside.

The Great Backyard Bird Count is not a contest, though. It's an attempt to document the distribution and numbers of birds at the end of winter throughout the continent before spring migration begins in March. A project of the Cornell Lab of Ornithology and the National Audubon Society, it combines high-tech web tools with observations of birds made by families, individuals, classrooms, and community groups who count the numbers and kinds of birds that visit their feeders, local parks, schoolyards, and other areas during any or all of the four count days. Then they enter their observations in a user-friendly, state-of-the-art web site.

In return, participants are rewarded by hourly updates on

their web site maps of the count as it progresses. Watching while lights blink on from the southern United States to Canada's Northwest Territories is fascinating. And each state also has its own map and statistics.

The folks running the survey aren't interested in extremely rare birds. Their focus is on the more common birds and in keeping them common. Even if we look for birds in bad weather, as I did, and don't see them, we report it. We also report snow depth, so scientists can continue to study the connections between weather patterns and bird movements. For instance, American robins have been staying farther north than usual in the winter. Is it a coincidence that there has been little or no snow in these places? Does global warming have something to do with it?

The Great Backyard Bird Count is a welcome diversion near the end of winter. As someone concerned about the welfare of wild birds, I am pleased that what provides entertainment for me is helpful to scientists trying to better understand birds and bird populations.

FEBRUARY 20. On this hazy, springlike afternoon, I glanced out our upstairs hall window and spotted a northern harrier dipping low over First Field. Running downstairs, I grabbed my binoculars and went outside to catch what I thought would be a brief look at the bird.

Instead I watched not one, but two immature harriers sailing back and forth over the field. They frequently hovered low, barely above the dead grasses, fluttered their wings and fanned their banded tails in an effort to harry their prey from hiding. They were likely in search of eastern meadow voles, one of several vole *(Microtus)* species that are their principal foods, because for the past two years our field has been overrun by them.

In addition to prey, the First Field offers ideal hunting perches for northern harriers—several dozen scattered locust posts driven into the ground to mark wild asparagus plants. As I watched, one of the harriers landed on a locust post to look around and preen itself. Then it flew off.

Because our autumn sightings had always been brief (once a harrier saw us, it would fly away), I assumed that the winter northern harriers would be equally shy. But they weren't.

First one bird returned to a locust pole, leaned down to pick at its feet, then plucked at its breast. Mostly, though, it looked around, its rusty-colored chest gleaming in the weak sunlight. The second bird landed on a catalpa tree in the field. The rest of the afternoon I watched as they alternately perched and preened on the posts or hunted over the grasses, gliding on their long wings, then flapping slowly, occasionally diving into the grasses, disappearing and reappearing like phantoms.

For hours I sat on the veranda in the warm, fifty-degree sunshine, my binoculars trained on one harrier or the other. Twice I thought they had continued on their way as they flew out of sight, but in a few minutes one would reappear on a locust pole, nonchalantly preening as if it had always been there. Finally, near dusk, one flew to a tree branch at the top of the field to sit and watch, and the other disappeared down the ridge.

Northern harriers are primarily birds of wetlands and prairies. Although they do breed occasionally in wet meadows and reclaimed surface mines in Pennsylvania, their breeding numbers are low.

A circumpolar species, northern harriers live in Europe and Asia as well as North America, where they breed across Canada and the northern United States. Their genus name *Circus* is Latin for "circle in the air," referring to their hunting technique, and their species' name *cyaneus* is Latin for "dark blue," the male's back color that is, in reality, dark gray. Formerly known as marsh hawks, other common names for northern harriers include "blue hawk," "mouse hawk" and "white-rumped hawk." The name "harrier" comes from the Old English word "herigian," meaning "to harass by hostile attacks."

Unlike most North American raptors, the female and male northern harriers have different colored plumages. The female—which is the larger sex—has a dark brown head, back, and upper wings, while the male's head is medium gray and its back and upper wings dark gray. Both have brown-streaked whitish breasts and that telltale white rump patch. Another peculiarity in northern harriers is their owl-like facial disks, which recent studies have shown focus sound down on their ear openings so they can locate prey by sound, just as owls do.

Northern harriers are opportunists, eating a wide variety of prey such as mice, rats, frogs, small snakes, lizards, crayfish, insects, small birds, and carrion. Because of the size difference between sexes, the smaller, more agile males catch a greater number of small birds, especially in brushy edge habitats, while the larger females prefer to hunt for mammals in open areas.

During a winter day northern harriers may fly a hundred miles in search of prey, but the size of their feeding territory ranges from forty acres to a square mile as they systematically quarter the same hunting area, just as I saw them doing on First Field today.

FEBRUARY 21. One of the harriers sat sunning itself on a locust pole as we ate breakfast. Then it took off, hunting back and forth over the field for a short time before it was gone.

Next, watching the bird feeders, I had a good view of two meadow voles rushing in and out of their network of tunnels in the melted mass of millet seeds and sunflower seed shells that blankets the area below the back steps. They flashed in and out of their holes like plump gray rockets.

Voles, the late researcher and writer Frances Hamerstrom discovered in her central Wisconsin research area, are the mainstay of northern harriers' diet. To emphasize the importance of voles to northern harriers she subtitled her book about northern harriers *The Hawk That Is Ruled by a Mouse*, meaning meadow voles, which are popularly known as meadow mice. More voles mean more harriers nesting. In that case, the old males will try to mate more than once and young males still in their immature plumage can also breed.

If the vole numbers are high, male harriers are not as monogamous as they are during low periods. And although the same harriers return year after year to the same marsh if they have raised young successfully there the previous year, they always find new mates.

Males first attract females by what Hamerstrom called "sky-dancing": "Swift twisting on silvery wings . . . [they] danced over sere meadows against lead-blue skies." They sky-dance during their migration north and also do it to advertise their territories.

The nest is built, mainly by the female, in heavy vegetation on the ground. Three to nine pale blue eggs are laid that turn dull white later in the thirty-one- to thirty-two-day incubation period. The female is the sole incubator and the male provides food for her and later for their nestlings.

Young harriers can crawl into the shade right after they hatch, and at two to four days of age they can crawl a yard. Although the male helps supply food for the young, only the female has the instinct to tear it up for them. If she dies, Hammerstrom noted, the male continues bringing in food, but he dumps it into the nest whole, and the young starve.

Harriers stoutly defend their nest from a variety of predators including red-tailed hawks, which sometimes kill and eat harrier young. They also harass crows because they are egg thieves. They even dive at foxes and white-tailed deer, the latter probably because they might step accidentally on nests.

People may also be dive-bombed if they stray too close to a nest.

After nearly five weeks in their nest, first the male nestlings fledge and then the females, but their parents continue to feed them until they migrate. Newly-fledged harriers cruise about half a mile from the nest site, exercising, playing, meeting their parents flying in with food, and mock-killing inanimate objects, but they don't hunt and their parents don't teach them to hunt.

Most young migrate approximately three weeks after they fledge, although some youngsters stay longer and even set up temporary ranges. Those that do migrate, migrate singly, and head for stopover areas where they set up temporary home ranges. The majority set their course for the southern United States or as far south as northern South America, but many stay in areas where the minimum winter temperature remains at ten degrees Fahrenheit or above. Often they roost on dry mounds on the ground, in groups of less than ten, and sometimes they also roost with short-eared owls.

Northern harriers are the most enjoyable of all the raptors I have watched on our mountain over the years, not only because they are easy to identify, but because their low-flying hunting techniques bring them down close. Certainly the beige grasses of First Field and the homely locust poles were enlivened yesterday and today and back in January when the northern harriers came to hunt.

FEBRUARY 22. A dusting of snow on the colder spots along Greenbrier Trail. I watched a sharp-shinned hawk try to grab a downy woodpecker as it landed on the side of a tree. The downy was merely indignant, flying off and calling loudly as it foraged on other nearby trees. After five minutes of protesting, tufted titmice flew into the area and the sharpie, having missed, moved on.

A few flakes fluttered down so I sought shelter beneath a

single large hemlock below Ten Springs Trail to write notes. In front of me a full complement of beige leaves, which were as translucent as Japanese screens, trembled on a pair of American beech trees.

Walking back up the road, I flushed a large bird that flew upstream and out of my sight. But as I slowly proceeded, still scanning the trees, the bird flew past me downstream and landed high in a tree. It was a great horned owl that sat erect, its ears perked up, and kept turning its head around as I watched it through my binoculars. When I lowered them because my glasses started to fog up, I couldn't relocate the owl with either my naked eyes or again with the binoculars no matter how closely I scanned the area where I thought the owl was. Either it had flown away in those seconds that I put my binoculars down or it blended so perfectly with the leafless tree that I could not see it. Talk about shape-shifting! How could the owl have disappeared so completely?

I was awakened abruptly at 11:05 P.M. by the sound of tundra swans calling low over the house. Spring migration has begun!

FEBRUARY 23. A damp, gray day—not the kind of weather for surprises. But I had two before 10:00 A.M.

The first occurred while we sat eating Saturday breakfast. Through the kitchen window I spotted large animals up in First Field that did not look like deer. They were turkeys—seven of them—foraging across the field as we watched. One was a male. He kept partially fanning his tail and rushing and/or circling around the three hen turkeys close to him. Had he already formed his harem? If so, he was at least a month earlier than usual. They continued on across the middle of the field and finally disappeared from sight behind the barn. What great entertainment they provided while we ate our cheese omelets and pumpkin muffins.

Later, as I was walking along the Laurel Ridge Trail be-

tween the Guesthouse and Short Way trails, I had my second
surprise. After topping a short rise, I looked down on an animal that was trotting along the trail, its long, bushy, dark tail
hanging down, its head close to the ground, its back toward
me. It was a gray fox. It never did see me as I stood watching
through my binoculars until it reached the powerline right-
of-way, turned and glanced down the trail toward me, and
then disappeared along the edge of the right-of-way. Who
cares if the sky is gray? On such days the wild animals seem
bolder and/or less wary. Such an unexpected sighting set me
up for the rest of the day.

As I neared the Far Field, I heard Canada geese. These
were definitely not locals. At least one hundred, flying in a
wavering wedge, headed north, and emitted "goose music"
as the late Aldo Leopold called their honking cries.

It warmed up to fifty-eight degrees with strong sun by
midafternoon so I sat out on the veranda. Suddenly, a female
pileated woodpecker yelled from the backyard. She had
landed on the staghorn sumac shrub, which swayed back and
forth from her weight, and she poked her beak into the red
fruit. Sumac is a favorite winter food of pileateds. Another
pileated called several times, no doubt her mate keeping in
touch, but she merely looked up and around, ducking her
head as if she did not want to give away her location, and continued eating almost furtively without answering him. He finally flew off. After about ten minutes of eating, she flew to a
nearby black walnut tree and then called loudly several times
before flying away. It seemed as if she had been practicing de-
ception on her mate.

FEBRUARY 24. What sound better signals the waning
of winter than the clear, musical "peter-peter" of the tufted
titmouse? Some years singing begins as early as mid-January,
and once I heard singing on a mild December day, but this
winter I heard the first song in late January during the Janu-

ary thaw, which is when the titmice usually begin practicing for the season to come.

Known as the "Peter bird" because of its distinctive song, the tufted titmouse is a bird of the eastern woodlands. Their song is "loud and persistent," in the words of ornithologist Arthur Cleveland Bent. "It is a joy to hear it tuning up in January, when so many other birds are silent." And, as I've discovered, whether or not they sing doesn't seem to be influenced by the weather. Time after time I have found them singing in the bitter cold, defying the winds, with the temperature as low as one degree below zero (Fahrenheit).

For many people, tufted titmice are winter birds because they come so easily to bird feeders. During the last fifty years they have extended their range northward, probably as a result of increased bird feeding as well as climate warming and the reversion of abandoned farmlands into the kind of mature hardwood forests these nonmigratory, mast-eating birds prefer.

Beechnuts are their favorite wild food, followed by acorns and smaller nuts, but they are perfectly satisfied with black-oil sunflower seeds from our feeders. Not only do titmice crack and eat them on location by holding them between their feet and pounding them open with their sharp, black bills, but they frequently snatch a seed and fly off with it so they can cache it for later use. Most of the time they shell it before caching it under loose tree bark or in the furrows, cracks, and rotted areas of trees. They have also been observed caching seeds on the ground and between two tree branches or at the end of broken tree branches and twigs.

When titmice are hungry, they are more likely to eat instead of hoard, and they always choose the heaviest sunflower seeds. They are also fruit and insect eaters and will glean bark crevices for insects throughout the year. Caterpillars are their specialty, making up as much as 38.3 percent of their diet according to one study, followed by mast (24.2 percent), which

---

they eat only between August and February. Pine seeds and the fruits of Virginia creeper, blueberry, hackberry, blackberry, bayberry, and mulberry are other popular choices.

Although they spend more than half of their foraging time in the winter searching for food on tree branches, twigs, and limbs, they also scratch through leaf litter looking for the mast leavings of squirrels. They are especially eager to find larger, broken acorns from red and white oaks and nuts such as black walnuts and hickory nuts, all of which are impossible for titmice to pound open. In our black-walnut–strewn yard, the squirrels leave behind bits and pieces of walnut meat for the titmice to glean.

When it is cold and windy, tufted titmice forage in sheltered areas such as overhanging stream banks, south-facing slopes, and close to the ground. Titmice also conserve energy by flying less and more slowly on cold days. They often sit on their feet and erect their feathers during the daytime. At night, their roosts in tree holes and cavities, they probably go into nocturnal hypothermia by reducing their body temperature ten to twelve degrees Fahrenheit like other members of the *Paridae* family do.

Tufted titmice are family-oriented birds, forming winter flocks that often consist solely of a mated male and female, joined perhaps by one or more of their offspring from the previous nesting season. Only occasionally do other, unrelated juveniles join the flock. These flocks start to form as early as August and are stable by the end of October. After that, if members die or disappear, they are not replaced, according to a two-year study conducted by Jeffrey D. Brawn and Fred B. Samson in a mature oak-hickory forest in central Missouri. They found that winter flock size ranged from two to five, although two—a male and female—was the most common size, but in Pennsylvania flocks averaged four titmice.

Males always dominate females and juveniles in these winter flocks. Usually the alpha male and dominant female are the ones that bred in the area the past season.

Operating under the peck-right dominance hierarchy first propounded by ecologist W. C. Allee, the dominant bird always has the right to peck subordinates. At our bird feeders, I can easily pick out the dominant titmouse. It is the one that chases off other titmice or takes their place.

The winter flock range is fifteen to twenty acres, and even when titmice join other bird species in mixed-species flock foraging, they stay within their own range, stopping at some invisible border as the other species cross it. In our mature, deciduous forest, tufted titmice associate most with black-capped chickadees even though titmice always dominate chickadees. Similarly, hairy woodpeckers dominate downy woodpeckers, downies dominate white-breasted nuthatches, and nuthatches dominate titmice. These relationships are easy to observe at our feeders and in our mixed-species flocks. In addition to the above-mentioned species, our titmice often associate with red-bellied woodpeckers, brown creepers, and golden-crowned kinglets.

As January segues into February, our titmice sing more frequently, probably triggered by increasing day length and higher testosterone levels. This leads to more aggression toward flock members. Gradually, spring dispersal occurs when the flock breaks up into pairs and single birds that claim their own two- to five-acre territory. Both paired and unpaired males sing from prominent perches in their territories, often meeting at a common border in what appears to me to be a singing contest.

Six days ago I watched a pair of titmice chasing and fluttering like butterflies in the wild grape vines. Today a titmouse landed in the front porch lilac bush, lowered its head, agitated its wings, and uttered a high-pitched, trilling call. I had observed what Donald W. Stokes and his wife, Lillian, call "wing-quiver," followed by "high-see call." This display is used in a variety of situations—when one male challenges another, before mating, by a female when her mate is feeding her, and by fledglings when they are being fed. Since I saw no mate-feeding or mating, I had probably witnessed a male aggression display.

Because pair-bonding can occur at any time of the year with titmice, there are no courtship displays other than possibly chasing. But once they are paired they stay together, and the male often feeds his mate. Copulation is always accompanied by the "high-see call," and the female and sometimes the male "wing-quiver."

Together the pair explore possible nest cavities in large woodland trees three to ninety feet from the ground. They do not excavate their own holes but depend on those previously excavated and abandoned by downy, hairy, red-headed, or pileated woodpeckers or northern flickers. They also readily take to nest boxes and sometimes hollow fence posts.

When the cavity is chosen, the female builds the nest, occasionally assisted by the male bringing nesting material. She starts with leaves, moss, bark, and sometimes feathers. Then she lines the nest with fine hair from such sources as raccoons, opossums, dogs, fox or red squirrels, rabbits, cows, cats, woodchucks, horses, and even human beings.

Once on a cold, windy day in mid-April, I found two large raccoons sleeping high up in a tree. In a few minutes, one titmouse, then a second one, landed on them. I watched as they snatched hair from the raccoons and flew off. The raccoons never even stirred.

It takes the female about four days to build the nest, which is finished, in most areas, by the end of April. Wing-quivering and mate-feeding intensify as the pair copulate frequently near the nest site. During this period she lays three to nine white eggs speckled with brown, purple, or chestnut-red dots.

Then the female incubates the eggs as the male continues mate-feeding both when she is on and off the nest. After twelve to fourteen days the eggs hatch. While both feed the young, the female also broods them for several days, so the male brings the bulk of the food during that period. They may also be assisted by a helper, usually offspring from the previous year that had remained with one parent or both throughout the winter. The nestlings fledge in seventeen or eighteen days and start feeding themselves five days later.

For more than a month, families stay together, and I frequently encounter them in July in our woods. Once in the Far Field Thicket, I watched five titmice, three of which were immatures. One kept trying to extract a piece of leaf from a cobweb, probably thinking that it was good to eat. On another July day, I found a mixed flock of mature and immature black-capped chickadees and tufted titmice traveling together, so even in summer these closely related species associate with each other.

By the end of August, most titmice young have dispersed and are looking for winter flocks to join. It is then that they appear in my backyard again. Hearing and seeing them, I realize that summer is fast waning. Once again I welcome back those sleek, dove-gray crested birds with white breasts as constant companions on my woodland walks.

FEBRUARY 25. Mostly clear and warming up rapidly. The woods were filled with chasing chipmunks. As I walked back up our road, chipmunks zipped across it and up the slope. I sat down to watch at least ten of them as they rushed back and forth across the road or froze on a fallen tree or stopped to scrub their faces or squealed and chased off other chipmunks.

At first it was difficult to make sense of what I was seeing because the chipmunks moved so fast. But I noticed that male chipmunks were tracking down a receptive female by sniffing along her path. Time and again she fought off her pursuers. Then she disappeared into a burrow and three males dove into it after her.

Suddenly she emerged and there was more frantic chasing,

squealing, and chirping. Once she leaped into the air, spun around, landed on her feet, and ran the opposite way to escape her pursuers.

I was watching what scientists call "mating bouts," when males chase a female in heat who outmaneuvers them time after time. I never did see any mating, but chases can last between two and nine hours. As researcher Lawrence Wishner concludes in his book, *Eastern Chipmunks: Secrets of Their Solitary Lives,* "The female's strategy seems to be to lead the males on a merry chase and eventually surrender to the one who is able to keep up with her. Presumably he is the cleverest and the strongest."

FEBRUARY 26. Another mild day. Along Greenbrier Trail, a porcupine was high in a branch of a triple-trunked sugar maple tree, curled up and sleeping and looking like a giant pincushion. The branch had been debarked in two small patches.

I still see porcupines frequently in winter and early spring walks, and whenever there is snow on the ground, their tracks meander along Sapsucker Ridge toward and into the spruce grove. But their numbers have declined from their high point in the mid- to late 1990s. Then they were all over the mountain. For instance, back in 1994, according to my journal, it was two degrees below zero as I snowshoed along the Far Field Road on February 26. At first there was no sign of any animal in the bitter cold, but then a movement far below caught my attention. A small porcupine was climbing around in a crooked tree. Choosing what seemed to be a dangerously thin branch, it first moved out on it in an upright position, reminding me of a small, performing bear in a circus.

Suddenly it gripped the branch with its four legs and tail, swung below the branch, and hung on like a three-toed sloth. Slowly it worked its way back and forth along the branch, gnawing vigorously on the bark and using its tail as both a guide and brace. The pencil-thin branch shook with every movement, and I expected it to break at any moment. Knowing that porcupines do occasionally die from falls, I was certain I was about to witness another fatality. Instead, the porcupine continued moving along the branch, still swinging and holding on with its legs and tail. Finally it returned to the tree trunk. Hanging from the branch by its hind legs and bracing with its tail, it groped for a hold on the trunk with its two front legs as it pulled itself to an upright position. From there it climbed a few feet up the trunk to a sturdier branch, where it stood on its hind legs to reach and gnaw on a smaller branch above.

In January and February 1997 at least five different porcupines foraged here during the daytime, even though they are primarily nocturnal feeders. Two of those porcupines, a large one and a small one, ate twigs, buds, and the needles of several Norway spruces in our grove. The porcupines were no more than ten feet off the ground and usually continued eating while I watched them.

Once there was snow on the ground I could easily chart their progress through the two-acre grove of spruces, because their distinctive, flat-footed tracks, like those of a small bear, are easy to identify. After they carve out a trail system they tend to reuse it, so it quickly becomes hard-packed, deep furrows in the snow.

That same small porcupine (or perhaps another one, because it's impossible to tell them apart without marking them in some way) also ate the bark of several small trees along the Far Field Road. One day it was in an American hop hornbeam. Another day it feasted on sugar maple bark. The following day it sat in a red maple tree and gnawed the bark of a wild grapevine that snaked over and around the maple tree's branches. Wild grapevine bark has consistently been a favorite porcupine winter food here.

Another favorite is the bark of chestnut oak, particularly

the highest, thinnest branches of two of our largest chestnut oak trees. Beginning on February 8, 1997, a large porcupine could be found most days in the chestnut oak tree fifty feet from where I had my winter stump seat. Snowy day after snowy day I swept off the inch or two of fresh snow and sat down on my hot seat to watch the animal.

One morning I arrived there before the porcupine. After cleaning off the two new inches of fluffy snow from the log, I settled down. Fifteen minutes later I heard a slight sound to my left. The large porcupine had plowed its way silently through the snow from its rock den on the far side of Sapsucker Ridge to the chestnut oak tree and was starting to climb up it. First it raised its four-clawed forefeet to grip the bark. Then it hoisted up its hind feet and rear end, an awkward ascent at best. Once it stopped in a tree crotch to shake off snow, but it was headed for the crown of the tree to resume eating the bark of smaller branches. It jockeyed about for the best position and held tightly to the thin branches with its forefeet. Finally, it started chewing on a small branch.

Porcupines prefer to eat the bark of the thin outer branches of deciduous trees because they can be eaten in their entirety, unlike larger branches that must first be shaved of their dead, thick bark layer. Tree bark and evergreen needles, incidentally, are not ideal foods. They produce only 2 to 8 percent of the nitrogen needed to build proteins, so porcupines lose weight all winter. Only by eating tree buds, leaves, and fruit in the spring and summer, and acorns and beechnuts in the fall can they put on enough weight to survive their nutrient-poor winter diet.

Porcupines usually choose a couple of favorite food trees near their winter den sites. Dens can be live hollow trees, rock crevices, hollow logs, or human outbuildings. They use the same dens year after year and many, especially males, move about every twenty-three days from one to another throughout the winter.

Over the years along Sapsucker Ridge I have found several hollow den trees, which are easily identified by the piles of brown, crescent-shaped droppings at their bases. I have tracked other porcupines to the large rock slide areas near the crest of the ridge. All of these den sites were occupied for only a few weeks at a time. But during two winters, the same porcupine lived under Margaret's abandoned house and fed exclusively on a large, front-yard hemlock. Today the hemlock is dead. However, all our other porcupine-damaged trees have only a few debarked or deneedled branches, probably because the porcupines moved from den to den and tree to tree throughout the winter.

For instance, after several weeks, the chestnut oak porcupines (and there were two of them because one day I found two large porcupines feeding within ten feet of each other) moved to a second large chestnut oak about a hundred feet from the first. The Norway spruce porcupines each worked three or four spruces at the edge of the grove, and while those trees were thinned, they were by no means killed. That same winter, down in the hollow, two porcupines ate either hemlock or basswood trees, both favorite winter food trees according to Uldis Roze in his excellent book *The North American Porcupine*.

Roze is a true fan of porcupines, after studying them for many years, and knows that they are despised for several reasons. One is their pruning or sometimes killing of trees.

Another is their craving for salt. Yet females especially need it to nourish their fetuses and, later, themselves because they lose sodium in their milk. Natural sources of salt are high sodium aquatic plants such as yellow pond lilies, the leaves of arrowheads, and aquatic liverworts. They also get salt from the bones of dead animals. But sometimes they gnaw on road-salt–impregnated tires as well as human-owned objects impregnated with our own salty sweat.

Third is their chief defense mechanism—thirty thousand

barbed quills. But Roze points out that they have three warning signals that predators should heed. Those with good eyesight should be wary of their black-and-white chevron marking, visible from behind, which means danger. Those with good hearing should listen for quiet, ominous teeth chattering. And those with a strong sense of smell should be warned by a pungent odor porcupines emit from their rosette, a bare patch above their tails. When all warnings fail and the animal attacks, the predator gets a face full of quills.

Even Roze, while trying to catch a seventeen-pound male in a beech tree, received his share of quills, including one that burrowed into his upper arm. After two days of pain, it emerged intact. "It looked beautifully clean, scoured by the body to a glistening freshness," Roze writes. "Like some mole of the flesh, it had ratcheted its way past muscles, nerves and blood vessels to emerge far from its site of entry."

Ever the scientist, Roze's experience led to a new discovery about porcupine quills. They are covered with antibiotics, a grease layer of fatty acids that inhibit the growth of harmful bacteria. No other mammal species is known to share this trait, so he postulates that it saves porcupines from being infected by their own quills when they fall out of trees.

Although quills cover every part of porcupines except for their underparts, muzzles, and ears, the short black quills on the upper surface of their tails are far more dangerous than the longer ones on their backs and necks. The powerful tail muscles can drive them so deeply into an animal that they disappear under the skin, but those from the rest of their bodies are usually only loosely attached to attacking animals and can be easily pulled out.

Roze also says that it is a myth that quills are hollow, so if they get embedded in flesh they should be cut off to allow air to escape and shrink the quills. On the contrary, the quills are filled with a spongy matrix and cutting them off may mean that they will be lost in the bodies of attacking animals. Because they can travel an inch a day, they could reach and pierce vital organs.

In late March, when the days remain above freezing, porcupines change their diet from tree bark to buds and leaves, both of which are high in protein. By the time females are ready to give birth in May or June, they are in prime condition.

Each female bears a single, precocial porcupette. Within half an hour its quills stiffen. Born with some teeth, at two weeks of age it can eat green plants, although it continues to nurse until it is four months old. Two months later, it leaves its mother, but it takes another year before the youngster reaches sexual maturity.

In October or November, a female leaves an odor trail as a signal to male porcupines that she will soon be entering her eight- to twelve-hour mating period. Sometimes several males arrive and fight, while she screams to keep the males away until she is ready. The male victor then guards her from others and performs what some researchers call a "three-legged dance," approaching her on his hind legs and tail while whining and grunting. Finally he sprays her thoroughly with urine. If she is not receptive, she shakes herself and walks away. If she is, she lifts her bottom and curves her tail back. After many short copulations, the male wanders off in search of other mates.

For a confirmed porcupine watcher like me, though, late winter and early spring are the best of times. It is then possible to see porcupines easily. Not only are the trees still bare, but the normally nocturnal porcupines often spend their days as well as their nights there, eating and sleeping in the treetops, just as the one did that I saw today.

FEBRUARY 27. Overcast and a misty rain at dawn. Lying in bed, I could hear water rushing down the drainage ditches. Every spring on the mountain spouted water into our stream.

As I went into the kitchen, the intercom buzzer from our guesthouse went off. Our son Steve, who was visiting for a few days, yelled, "Mom, come quick! There's a beaver in the stream below the guesthouse."

At first I didn't believe him. But as he insisted, I grabbed my binoculars, pulled on my boots and jacket, and ran down in time to see an adult beaver emerge from the culvert pipe beneath the road. I was amazed at how large it was, especially when it stood up on its hind legs beside the drainage ditch to look around. We had plenty of time to study its paddle-shaped tail lying flat on the lawn and admire its sleek, dark brown coat.

Although Dave and Steve stood with me on the guest-house porch fewer than fifty feet from the beaver, quietly talk-ing, it seemed supremely unconcerned by us. Perhaps it was looking over the terrain and trying to decide if it had poten-tial as a future home. But two houses and three adult humans were probably enough to discourage it. After five minutes of apparent indecision, it continued up the drainage ditch to-ward the powerline right-of-way, wading through six inches of flowing water.

I rushed back to our house to rouse Bruce. Together we ran up our driveway to the powerline right-of-way ahead of the beaver. While I stood on one side, scanning downstream with my binoculars, Bruce crossed the ditch and set up his camera and tripod on the embankment above.

When the beaver came into view, I called quietly to Bruce, "Here it comes."

Remaining still and out of sight, I watched while the beaver attempted to scale a fallen tree and then toppled over backward. Undeterred by that setback, it tried again and sur-mounted what must have been the last of dozens of fallen trees that span the stream.

It waddled determinedly up the ditch, by then only inter-mittently filled with runoff water. Finally, it sensed Bruce above it and stopped. Again it sat up on its hind legs and peered toward Bruce, who shot picture after picture before the beaver slowly turned around and headed downstream. Again we watched from the guesthouse porch as it went down into the culvert pipe and emerged in the stream directly below us. Hopeful that it might set up housekeeping in our marshy meadow, I didn't follow it down the mountain. But our marsh is only an acre at most, and it doesn't have enough of the preferred winter food trees—aspen, sugar maple, tulip poplar, and willow—or the aquatic plants, forbs, and grasses that beavers eat in the summer. They also like flat terrain or valleys and large streams with enough water for damming.

Probably the beaver we saw was a two-and-a-half-year-old that had voluntarily left its parents' lodge and was looking for a home of its own. But according to the books I checked, it was a couple of months ahead of schedule. Usually a mature beaver leaves its parents and younger siblings between April and September and becomes a floater segment of the population, following water courses as far as twelve-and-a-half miles from its natal home in search of its own turf. But this beaver had instead set out on its own during February's thaw, perhaps convinced by the sound of running water that spring was here to stay.

I took a walk down Ten Springs Trail and up the road later in the morning and did not get a glimpse of the beaver, so it probably headed back down to the Little Juniata River to explore other stream systems.

The first turkey vulture rocked past at noon and Bruce and Steve later counted twenty turkey vultures during an afternoon walk to the Far Field. A pair of eastern bluebirds also appeared as we sat out on the veranda, serenaded by nonstop song sparrow singing and the occasional calls of northern flickers, killdeer, and red-winged blackbirds.

Late in the afternoon, driving rain sent me indoors for the rest of the day.

FEBRUARY 28. The American tree sparrows are singing. Unlike their bell-like feeding calls that I hear all winter long, I never heard their high, sweet, clear songs that trend downscale until 1995 and then I didn't know what species of bird was singing. I called it the "mysterious singer" that sang deep in the dense blackberry tangle along Greenbrier Trail early in the morning. Day after day I tried to track down the singer, once flushing a flock of white-throated sparrows and golden-crowned kinglets in early April, but I never found it.

The following March I wrote, "The mysterious singer from early last spring is either an American tree sparrow or a dark-eyed junco. It sang today hidden in the forsythia bush near the garage and both species flew out as I walked past." I knew that all the bird books claim that the juncos' only song is a trill, which I hear every year at this time, so I was almost certain that my mysterious singer was a tree sparrow. A gift of birdsong tapes verified my supposition. Now I look forward to hearing at least some tree sparrow song in late winter before they head north to their breeding grounds, where they sing brilliantly and continually for about four weeks. Just as their musical feeding calls enhance my winters, so do their ringing songs signal the imminence of spring.

Pennsylvania is in the central part of their wintering range, but soon these so-called "winter chippies" will be replaced by their look-alike relatives, chipping sparrows. Their rusty caps resemble the caps of chipping sparrows, but they have a black spot on their breast that distinguishes them from their plain-breasted cousins. They also look like the European chestnut-capped tree sparrow, which is why American tree sparrows, which actually prefer scrubby edges of marshes, fields, hedgerows, and fallow fields in winter, and brushy willows and birches around pools and boggy meadows on their northern breeding grounds, were misnamed "tree" sparrows. They should probably be called "brush sparrows" instead, according to A. Marguerite Baumgartner, the first ornithologist to study them both on their breeding and wintering grounds.

Back in 1933 and '34, Baumgartner spent her summers near Churchill, Manitoba, observing the nesting habits of American tree sparrows. They nest "in the wet, brushy wastes of the Canadian Northland, beyond arable land or usable timber, from the northern third of the Hudsonian spruce timber as far north as there is any scrubby growth," Baumgartner wrote.

She also watched tree sparrow courtship, which depends heavily on song. Both sexes are aggressive and chase each other. During her two seasons in the North, she found twenty-six nests, nine of which she kept under constant surveillance, often watching throughout the night.

The nests were difficult to find because they are hidden in the densest tangles of scrubby thickets and most are on the ground. The female builds the nest, usually lining it with ptarmigan feathers, but they sometimes substituted whatever was available such as pintail feathers, dog hairs, lemming fur, or mosses. One even used scraps of waste cloth she found near the railway station.

Nest building is a leisurely practice, taking at least seven days, because, "after three or four trips for material, the pair usually flits off to feed about the marshy edges of a pool, and the observer might sit and shiver for two hours or the rest of the day before they return," Baumgartner wrote.

The female tree sparrow lays from three to six ovate, slightly glossy, pale-bluish or pale greenish-white eggs heavily spotted with brown. She incubates them from twelve to thirteen days but the male visits her and they go off and feed together. After the eggs hatch, they both feed the young for the nine days they are in the nest and for two weeks after they fledge.

Tree sparrows, Baumgartner discovered, have a strong homing instinct and return to the same banding station (or backyard bird feeder) in the fall. They migrate at night, mostly in same-species flocks, but sometimes with other species as well, and leave the North during the last two weeks in September. By early November they reach our mountain.

During the winter they eat about a quarter of an ounce of seed a day and prefer grass and weed seeds, such as pigweed, ragweed, lamb's quarters, crab and yard grasses, timothy, knotweed, bindweed, smartweed, and goldenrod. At feeders they particularly like millet seed followed by black-oil sunflower seeds. They also consume a wide variety of insects, their eggs, and larvae. On cold winter days I often watch them beat dried weeds with their wings to release seeds that they then pluck from the ground. Berries and catkins from trees and shrubs are other favorites.

As they feed, both sexes emit what naturalist/writer Henry David Thoreau described as their "tinkle of icicles" call. But late in February the males begin "spring tune-up and by mid-March the hedgerows ring with true summer song," Baumgartner observed. Just as they did today.

March

Roaming the woods and hills is a physical form
of identification with the unfolding, collective life
of our home on earth.

John Elder, *Reading the Mountains of Home*

MARCH 1. An American robin and song sparrows caroled at dawn. Along the Greenbrier Trail I watched wedge after wedge of Canada geese flying high and heading north. From the bench blind, I had a front-seat view of migrating geese. Hundreds of them continued streaming across the sky. Most groups contained 125 or more and they moved so fast I lost count. Some were far out over the Little Juniata River gap and I saw but did not hear them. I estimate that well over a thousand passed within my view, clearly fleeing north ahead of the predicted cold weather.

Once again the mammals have returned in the evenings to feed on birdseed. First a single fat raccoon. Then an opossum. Then two raccoons and the opossum. Tonight we watched the opossum face off with the fat raccoon. The opossum kept backing away and circling as the raccoon had its head down to the food, moving, but never looking up as if it didn't know the opossum was there. The opossum, on the other hand, was clearly worried but not frightened enough to move away, or maybe it was too starved to care. Finally, it relaxed and fed several feet away from the raccoon.

MARCH 2. At 6:30 A.M. an eastern bluebird woke me with his "tru-ly" song outside my bedroom window and later sang unceasingly in the yard even though it was eighteen degrees outside. The white-breasted nuthatches and tufted titmice were also in a tizzy. Chipmunks were still out and about in the woods despite the cold.

Along Sapsucker Ridge Trail, among fallen trees, chipmunks chased, up to a dozen at times. Up and down trees, along logs, even right past my feet they ran, mostly silently, but occasionally cheeping. A rotted log seemed to be a special place to be defended by one chipmunk (probably a female) who bowled over any other that came near. They appeared oblivious to me. At times the woods churned with chasing chipmunks expending incredible energy in their mating bouts.

Back on the home grounds, the male bluebird still rehearsed his one-song repertoire while a female sat nonchalantly on the telephone line. When he showed off the nest box, singing all the while, she looked bored.

A white-throated sparrow rendered his "poor Sam Peabody" song even as a wind picked up and clouds moved in from the west. I sat on the veranda at noon, enjoying the day, while American crows "cawed" over the field in dark formations. Suddenly their calls intensified as a huge bird flapped low over our yard. I grabbed my binoculars and ran after it. White above the tail but much too large for a northern harrier. White on its wing primaries from below, the tips of its wings curved upward—it was an immature golden eagle. It circled over Margaret's Woods as the resident red-tailed hawk checked it out, looking small beside the enormous eagle. Then the sky clouded over and by 1:00 P.M. it was snowing and blowing in a brief flurry.

I switched on the back porch light at 8:30 P.M. and inter-rupted quite a drama. The triplets were back and so was the skunk. The larger raccoons moved menacingly toward the skunk as it tried to feed on the steps. Instantly its tail shot up and it stamped its front foot, then turned away. We held our breaths. But instead of letting loose, it made a dignified re-treat, its tail still high in the air. We couldn't figure out if the raccoons were at all intimidated by its threat of the ultimate chemical weapon, but when the skunk quietly returned, a few minutes later, its tail was down and it fed off to the side just above the outside cellar door. All of them were feeding peace-fully when we went up to take our showers. Later, when I opened my bedroom windows, there was not even a faint whiff of skunk.

MARCH 3. A Carolina wren fed on birdseed this morn-ing, the first time this winter. March is always the starving time for wildlife. Often the nocturnal creatures are abroad even in the day time. The trails are littered with coyote and fox scat. This easy winter has been a boon to both predator and prey species.

Not so the winter of 1994. That was the killer winter—the coldest and snowiest on record. I snowshoed most days from the beginning of January to near the end of March. It was also the winter that Bruce decided to buy a secondhand bull-dozer. He followed that purchase the following winter, which was also snowy, with a new tractor equipped with a front-end snowblower.

He also started taking off winters from his job, because he had so many nights in 1994 when he arrived at the bottom of our road at 11:00 P.M. and then had to snowshoe the mile and a half up the mountain because he couldn't get the car through the drifts. Weeks before our pickup truck plow had been defeated by the unending snowstorms. Three separate times we paid a bulldozer operator to clear the road. But each time we had to wait several days before the operator had time to open up our road to the outside world. In the meantime, Bruce and Dave had to pack in food on their backs.

Since then, of course, the winters have been ridiculously easy and the road snow-free longer than ever, sometimes even in the middle of February. The hard winters in the early and mid-1990s seem gone forever as global warming kicks in with a vengeance.

But on this date in 1994 it was thirty-two degrees and snowing hard. It had been snowing for more than twenty-four hours, giving us eighteen new inches of snow on top of what was already there—a grand total of thirty-eight inches of standing snow. Penn State University was closed and Bruce and Dave snowshoed down the mountain to dig out our car in the afternoon. Winter and white and snowshoeing seemed permanent conditions, though, and the idea of spring and birds and seeing the bare earth again an impossible dream. Indeed, more snowstorms followed and I was still on snowshoes the last week of March. An enormous avalanche of frozen ice near the bottom of the hollow didn't melt until late April.

But tonight, in this first year of the twenty-first century, the sky was clear and starlit. Bruce set up the telescope and we looked at the moons of Jupiter, the Orion nebulae, and the Pleiades star cluster. The latter was the most spectacular—stars without end and our earth such a minuscule portion of the vast universe. Every human should feel insignificant after a night of stargazing and, at the same time, filled with wonder at the immensity of space. If only all self-important politicians were forced to look at the stars.

While we were busy admiring the universe in the front yard, a raccoon and opossum, at opposite ends of the feeder area, were attending to the practical necessity of life on earth—filling their bellies. In the distance, the bass "who-who" of a male great horned owl reverberated off the ridges.

MARCH 4. The dawn chorus of song sparrows, Carolina wrens, and northern cardinals seemed to be welcoming the mostly clear and warming day. A ruffed grouse drummed as I walked along Laurel Ridge Trail. When sunbathing against the Far Field road bank, I had to take off my jacket. Amazing how warm the ground is at this time of year even though the air is cool. This is the month for light-worshipping as the sun pours down unimpeded by tree leaves. Tree branches glisten and laurel leaves shine while the dried field grasses glow a beigy-pink. All reflect the light until the whole mountain shimmers on sunny days.

On top of Sapsucker Ridge, there was a skim of ice on four of the ephemeral ponds. The shredded bark of a fallen striped maple tree along Sapsucker Ridge Trail looked like deer work, but the gnawing of half the bark from the base of a medium-sized beech looked more like porcupine.

A northern cardinal and a tufted titmouse duetted back and forth. "Petty, pretty, pretty"—"peter, peter, peter." Then, for a few minutes the cardinal prevailed, singing seven or more "prettys" over one weak "peter." But later the titmouse seemed to get his second wind and delivered a steady barrage of "peters" after each cardinal round. Then a second cardinal joined in with "cheer, cheer, cheer." That seemed to shut the titmouse up.

While I was making lunch, Dave buzzed the intercom to announce that a flock of tundra swans was flying over the mountain. I rushed out to look, but the white, wavering line of sixty or more was so high in the blue sky I could barely see them.

MARCH 5. The world is aglitter. The successive rain, sleet, and inch of wet snow that fell overnight put a sheen of incomparable beauty on the mountain. Hanging from each limb are frozen droplets, some longer than three inches and all catching beams of sunlight against an Alpine-blue sky.

Atop the limbs, branches, and mountain laurel leaves is at least an inch of snow. As I looked directly into the sun through the icy trees my eyes were dazzled by the crystalline light. When I gazed down First Field at the icy trees to the left on which the sun was shining from the side, they resembled the twinkle of miniature Christmas lights—blue, red, green, yellow, and orange combined with the translucent droplets sparkling in the tree branches.

The road to the Far Field was even lovelier and where the ground was exposed, steam rose almost as thickly as that at sulphur springs. To remind me that this was a spring and not a winter snow, I heard the constant calling of migrating Canada geese and it quickly warmed up enough to set the woods dripping. I kept stopping and looking, utterly awestruck at what I was seeing and hearing, and I wondered if ornithologist and writer Alexander Skutch was right when he postulated that we evolved so that some appreciation for the beauty of the earth would emanate from at least a few of our hearts.

"All this for me, Lord?" I reverently questioned as I continued being stunned by visions of beauty wherever I walked. I marveled at a small beech tree still retaining all its bleached beige autumn leaves in the midst of the surrounding white, green, and black of mountain laurel and bare-limbed hardwoods. And even the deer standing in our driveway as I walked toward them seemed like a vision, infusing vigorous life into the world of snow and ice.

At last I turned toward home, walking along the bottom edge of Sapsucker Ridge, and noticed that the treetop ice droplets still served as prisms for all the colors of the rainbow although the preponderant lights were red, orange, and yellow with occasional flashes of pale green.

By afternoon the ephemeral glory was gone.

MARCH 6. Birds flocked to the feeder area including two American crows, the first I have ever had at our feeders.

They scratched the wet snow aside and scarfed up seeds as they waddled about. One kept looking up alertly and flew off and back several times, calling a warning to its bolder companion. That companion paid attention to nothing but eating. The other birds kept their distance, but continued flying in and out of the feeders above the crows. As we looked from the kitchen door window, we had our closest-ever chance to examine the blackness of crows and their large, heavy bills that looked capable of cracking black walnuts.

I walked the Greenbrier Trail and heard an eastern towhee repeating its name—an unbelievably early return date for that species. Northern cardinals sang and red-bellied woodpeckers "chiv-chived." Four deer scattered awkwardly from a side hollow, moving through the fog like phantoms. Fog billowed up the hollow and swirled through the trees, the kind of day that looks uninviting from inside the house but inspiring once I go out.

From Greenbrier Trail I dropped down to Ten Springs Trail and heard a singing white-throated sparrow and the "dee-dee-dee" of a killdeer as it flew over the mountain. Following a deer track across the steepest part of the hollow, I slowly picked my way to a rivulet that I followed to our road.

Walking back up, I spotted a small, green caterpillar with a light dot-and-dash line on its back crawling along a blackberry cane. Half of its body was on a swelling bud and I saw a thin line of webbing strung from thorn to thorn up the cane. As I watched, it moved its head searchingly over the bud. Then it slowly turned and headed back down the cane. It seemed awfully cold and early for a caterpillar to be abroad.

By late afternoon it had warmed up and I sat on the veranda, enjoying the warm wind and intermittent sunshine, and watched a turkey vulture sail past. The constant "tick-tick" of a scolding songbird pulled me up out of my chair to investigate. I thought I glimpsed an American kestrel out beyond the barn.

Sure enough, the male kestrel was back. He flew first to the tip of a catalpa tree in the field and then perched on the wire above the kestrel nest box.

In the meantime, more and more American crows gathered above Laurel Ridge—close to forty—cawing, soaring in the wind, and then flying over to Sapsucker Ridge. All the while the kestrel sat tight.

Later, near dusk, he called "killi, killi, killi" from the top of his favorite power pole, duetting with a couple of calling robins whose "tut-tut-tuts" sounded remarkably like his calls.

MARCH 7. Again I sat on our veranda late in the afternoon. It was fifty-seven degrees, the March wind was blowing, and the yard birds were calling, singing, and chasing.

A pair of red-tailed hawks, sounding to my ears like tinhorns, emitted what the experts refer to as their "chwirk calls" while they circled above Sapsucker Ridge. Then one alighted on the remnants of an old nest and the other flew above it, circling, calling, and landing in nearby treetops, its legs extended downward. I was watching the so-called "talon-drop" display that redtails perform in courtship or to defend their territory. Because they eventually flew off together, instead of one routing the other, they were probably a mated pair and not two males fighting over territory.

Last March 24, the redtails used our First Field, yard, and wooded ridge tops for even more elaborate courtship rites. I had been away for most of the day, but what a homecoming I received. Dave announced that redtails had been courting for hours above the field. When I stepped outside to put on my hiking boots, the redtails flew low over our yard, "chwirking" and performing talon drop. Then the female flew to a tall white pine on the far side of our powerline right-of-way, and the male streaked after her.

Because of the distance, we were not positive that they mated, but the male did appear to land briefly on the female's back. According to ornithologists, the female leans forward with her wings dangling loosely at her sides, inviting the male to mate. Copulation lasts from five to ten seconds.

Eager to make the best of the beautiful day, I hiked along Sapsucker Ridge Trail. The redtails continued to fly up and down the ridge top, giving both their piercing whistles and "chwirk" calls. I sat against the largest oak tree on our property and watched as the male redtail sailed over First Field "chwirking." Then he flew straight down and up and then down again, like an airplane caught in extreme turbulence, before disappearing from view. Although this so-called "undulatory flight" is thought to be a territorial display, he appeared to be using it in courtship, because only a couple of seconds later the female skimmed through the woods seven feet from the ground and a mere thirty feet from where I sat undetected by the courting pair.

The male zipped past after her and both disappeared over the side of the mountain toward the valley. I followed their path on slow, earthbound feet, stumbling over rocks and fallen limbs. By the time I made it to the edge of the ridge top, they had disappeared for the day.

Redtails are faithful to each other and their territory for as long as they live. If one dies the survivor will hold on to the territory until another mate appears. The territory itself ranges from half a square mile to more than two square miles, depending on the abundance of food, nesting and perching sites. I suspect that they nest at the end of Sapsucker Ridge, because we usually see a redtail family flying and calling in the area in early summer.

Shortly after dark, Dave called us outside to listen to a steady, high-pitched "toot-toot-toot-toot" call that sounded, he said, like someone sharpening a saw. Steve had reported a similar call, near the old dump, when he was here last week, but the noise from the interstate was so strong that we didn't hear it. Steve said it was a northern saw-whet owl. After we listened, we went inside and put on the bird tape. Then we went outside a second time. The bird was still calling and there was no doubt that it was a saw-whet owl, the first time I have ever heard one call that I am aware of. On the other hand it may have been a case of someone alerting us to what we had previously ignored as an aberrant night sound.

MARCH 8. A breezy sixty-five degrees in hazy skies. Dave reported hearing the saw-whet owl calling at 5:30 A.M.

This is record-breaking warmth although so far there has been neither sight nor sound of eastern phoebes or wood frogs. Still determined to enjoy each day as it comes and not worry about whether or not this is a result of global warming, I sat on Turtle Bench in a t-shirt in midmorning.

Walking along Sapsucker Ridge Trail, I watched a pair of mourning cloak butterflies twirling in the warmth. Dave saw the first Compton tortoiseshell butterfly out around the guesthouse. These closely-related butterfly species in the genus *Nymphalis* hibernate in their adult stage and emerge early in spring to mate and lay eggs. To see them is as much a sign of spring as migrating birds.

The first field sparrow flew into the feeder area. Three migrating fox sparrows also stopped by. The yard reverberated with the trills of eighteen dark-eyed juncos. Watching feeder birds at noon, I noticed a mourning dove pair perched side by side on a tree branch, she wriggling her tail, he fluffing his feathers. Then he billed around her face for a few seconds before mating. After that, they faced each other on the branch, literally billing and cooing for a minute or more.

MARCH 9. Warm and overcast. The birds were calling and singing and down by our tiny pond I could hear the first ducklike quacks of male wood frogs.

As I walked along Greenbrier Trail, I was suddenly stopped in my tracks by growling shrieks that I thought, at first, were made by an owl being harassed by American crows. Yet the intermittent shrieks did not seem to be coming from the same place as the cawing crows.

Puzzled by the sound, I sat down in a dense grove of striped maple trees just as it started to shower. Under my umbrella, I continued listening to the growling shrieks, but I still could not pinpoint the source. When it stopped raining, I resumed my walk, moving quietly as I scanned the trees above the trail.

Finally, I spotted a raccoon climbing out on a tree branch and then down the trunk of a red oak in pursuit of another raccoon perched on a branch on the opposite side of the tree from where I was standing. I quickly decided that the shrieking raccoon was the female and the pursuer the male.

More shrieking erupted, and then the male climbed several feet above the motionless female and moved restlessly around on branches and tree trunk as if he were trying to figure out another approach. Slowly he slid down the trunk head-first toward the female, all the time emitting calls that sounded like clicking castanets, and settled into a tree crotch next to the female. He continued to be restless, periodically moving around in slow motion, while the female, partly hidden from me by a mass of grapevines, remained still.

Frustrated by my somewhat obstructed view of the proceedings, I climbed up the ridge so that I could look down at the raccoons' tree. I watched as the male descended below the female and then climbed back up above her. She shrieked several times, he growled, and finally he lay down on top of her. They looked like amorphous, fuzzy blobs in the tree, veiled, as they were, by the grapevines. Finally, he moved around, his tail up, then down, his forepaws stroking her face. Whenever he moved, she shrieked, but he was firmly planted on top of her.

After several minutes, the male raccoon arose and chased the female higher up the tree, she shrieking, he making purring clucks. They met nose to nose and patted each other's faces with their forepaws. Then the female curled up in a furry ball while the male again restlessly moved around, sniffing her backside as if he were checking to see if she was ready to mate. At last he retreated a couple of feet below her and settled down just as the sun appeared from behind the lowering clouds.

Because I had other commitments, I reluctantly turned homeward when the raccoons quieted down, convinced that I had seen only a portion of what seemed to be an intricate courtship. When I read up on the subject, I learned that raccoon courtship and mating are indeed intricate and variable. One researcher back in 1956 observed them mating in Kansas only once. After reading his account, I understand why no one goes into any detail about raccoon mating. If it were made into a movie, it would probably earn an R or maybe even an X rating.

A pair of researchers in Texas radio-monitored raccoons in 1990–92 and discovered an incredible variation in their mating strategies, even within the limited population they studied. Consortship, which they defined as a "diurnal association between an adult male and female observed resting together or sharing a small den structure with a single opening," lasted anywhere from one to three days. During 62 percent of consortships, one female consorted with only one male. The rest consorted with between two and four males. Those with shorter estrous periods, between two and three days, consorted with only one male. Those with longer estrous periods (four to six days) consorted with more than one male.

The males themselves formed loose groups in a home range, and the dominant male consorted with most of the females while subordinate males tried to find a female before the dominant one did. In addition, solitary males roamed from home range to home range in search of females.

Whatever their courtship and mating strategies may be,

though, the result is the birth of two to seven cubs sixty-three days later in the female's tree den. The cubs open their eyes at four weeks of age, when the female begins to wean them. At six to nine weeks old, she moves them from her den tree to a ground bed on the forest floor or in a wetland. By then they are playing exuberantly. A few weeks later they begin to accompany her on short, round-trip excursions. Within a week they are able to move and bed together, following their mother as she emits a constant low, grumbling purr. When they disobey, she slaps their rear ends. Once they are thoroughly weaned, at four months of age, they are more independent, trailing behind or ahead of her. Throughout the summer she teaches them to climb and hunt for food.

They spend the autumn fattening up on a wide variety of wild fruits and nuts, especially acorns, and usually den together in the winter like our triplets must have done. Raccoons are not true hibernators, but here they tend to spend the bitter months of mid-January through most of February in restless sleep. Their body temperature drops from 100.6 to 96 degrees Fahrenheit, they don't urinate or defecate, and they live on their stored body fat, losing half their body weight by spring.

Once they leave their winter dens, families disperse. Males travel farther than females, as much as ten miles or more, and one radio-tagged male went 26.7 miles. I wonder if the raccoons I watched today could have been the parents of the triplets?

MARCH 10. While I was still lying in bed at 5:30 A.M. I heard tundra swans going over. When I stepped outside an hour later, I counted another wedge of fifty-nine.

As I started up First Field Trail through the forest, I heard still more tundra swans calling. I raced back to the open powerline right-of-way for an unimpeded view of one hundred tundra swans passing directly overhead, their white wings gleaming against a deep blue sky, their black feet trailing out behind them, my own special angels of the sky. There was one major wedge of fifty, plus several small, wavering lines of between nine and twenty-one that preceded and followed the central flock. Several minutes after they passed, five more swans flew over in a straight line. The major noisemakers of the day—northern cardinals and American crows—were temporarily silenced by the tundra swans as if in tribute to their size, beauty, and wild calls, like peasants in awe of royalty.

Thinking the show was over, I continued my walk along the edge of the forest. Ten minutes later I heard another flock. This time I ran to the middle of First Field and counted 125 swans flying in uneven, shimmering lines across the sky. After two more minutes, seventy-five came into view, then sixty, ninety, seventy, fifteen, sixty-five, and thirty-two rolled in waves over the mountain. Clouds gathered, turning the sky gray, but still they came, about six hundred in all, even as snowflakes floated in the wind.

In less than half an hour the tundra swan extravaganza was over for the day. Long after they were gone, their wild cries echoed in my head, cries flung in defiance of the wind and cold, proclaiming the advent of spring despite the wintry chill.

Tundra swans spend their winters in estuaries along the Atlantic coast, especially in the Chesapeake Bay and North Carolina's Albemarle Sound. But by mid-February many of them have flown to the lower Susquehanna River Valley and Middle Creek Wildlife Management Area in Pennsylvania's Lancaster and Lebanon counties. This is a major staging area where they await the first break in the weather to head northwest across the state, passing directly over our mountain, to another major staging area, the marshes on the southern shore of Lake Erie in Ohio.

Some of the swans arrive in Pennsylvania as early as mid- to late December if the winter is mild, and numbers continue to build up throughout January and into February. By late February, ten to twenty thousand are in Pennsylvania. This so-called eastern population breeds primarily in the northern tundra regions of the North Slope of Alaska, the Yukon, and the Northwest Territories and migrates southeast across the continent.

Both during migration and on their nesting grounds, the domestic life of tundra swans—unlike raccoons—is exemplary from a human standpoint. Not only do they mate for life, but they care for their young throughout their first fall and spring migrations. In addition, they are faithful to the same breeding territories and wintering sites. Even though they leave southern Pennsylvania by early to late March, it is early to late May before they arrive at their breeding grounds. There they defend their half-mile territory from others by singly or jointly chasing intruders.

Both parents construct their nest mound near water, using grasses, sedges, lichens, mosses, and sod that they pull out in the vicinity of the nest site. Often they reuse a previous nest but keep adding to it throughout egg-laying and incubation. The parents share incubation and egg-turning duties, after the female lays an average of four to five creamy white eggs, although the female spends 70 percent of her time on the nest in contrast to the male's 25 percent.

They both protect the eggs from a long list of predators—red and arctic foxes, golden eagles, parasitic jaegers, glaucous gulls, pomarine jaegers, common ravens, and brown bears—because if the parents lose a clutch, they do not renest. If they are successful in defending their eggs, the eggs hatch after a month, and downy, light gray, precocial cygnets emerge, ready in twelve hours to follow their parents to feeding areas in lakes, ponds, and pools. There they eat the seeds, stems, roots, and tubers of aquatic vegetation. Because of their initial buoyancy, the cygnets cannot dive, tip-up, or forage well under water for at least two weeks, so their parents often paddle submerged vegetation to the water's surface for them to eat.

In late September family groups form migrating flocks and begin the long migration, arriving in the Devils Lake area of North Dakota and the upper Mississippi River in Minnesota in early October. After feeding and resting for several weeks, they migrate nonstop to the East Coast, flying at about fifty miles an hour, and arrive from mid-November to mid-December.

I have heard them here as early as October 31 and as late as November 22. Until last spring their earliest northern migration date was March 8 and the latest March 31. But last winter the balmy temperatures of late February set them off, and they were migrating over the mountain by the hundreds on the first, second, and third of March.

I always hear them before I see them. High-pitched, musical laughter emanates from above the clouds, the sound that gave them their former popular name "whistling" swans. Nothing can compare to the sight and sound of tundra swans heading northwest, long-awaited harbingers of spring, as entrancing to me as "goose music" was to Aldo Leopold more than half a century ago, when Canada geese were a rare and cherished sight. Now that so many geese have become pesky residents in the valleys and apt to fly low over the mountain at any time of the year, instead of Leopold's totemic geese, I listen for "swan music."

MARCH 11. Warm, overcast, and humid, a misty rain falling off and on this Sunday morning. I went outside to sniff the air at 9:30 A.M., still clad in my nightgown and bathrobe. In First Field near the first rise I caught a movement that turned out to be a red fox hunting for meadow voles.

I slipped back inside to alert Bruce and to grab my binoculars and out we went to watch quietly from the veranda. When the fox disappeared over the rise we slowly walked across the lawn and up the driveway past the garage when again it came into sight, continually pouncing, pirouetting, searching, but never looking in our direction. Once it caught something and quickly swallowed it.

Slowly it worked its way up the field, sometimes hidden by the rise, sometimes not. This fox had a lot of gray and white on its coat and the only red it had was on its back. Even its tail was mostly gray and black with the characteristic white tip of a red fox's as well as its usual black stockings. It was probably the same cross fox we had watched back in January.

Often the fox looked in our direction but not at us so we had many wonderful chances to study its foxy face. We stood still for a long time on the road above the garage, the wind blowing our scent away from the fox so it had no reason to suspect we were there. Eventually it disappeared over the rise again and did not reappear.

We headed up the field along the path in the general direction we had last seen it, and coming up over another rise, we suddenly spotted it directly below us, the closest we had yet been to the fox. We froze, but it continued hunting, and we watched it catch another creature. Then it trotted purposefully down over the amphitheatre while we stayed where we were. As it emerged again on the top of the rise, we realized that the wind would now be blowing our scent in its direction. Sure enough, the fox stopped, looked at us, and then sped to the base of Sapsucker Ridge below the grape thickets. It paused again, glanced once more in our direction, and finally disappeared into the underbrush. We had watched the fox for almost an hour, the longest look either of us had ever had of a hunting red fox. I was impressed by its intentness. Unlike deer, which are constantly glancing up to look for possible trouble, the fox seemed totally preoccupied by hunting.

In the early afternoon I walked my usual circuit, and on Laurel Ridge Trail, before I reached Short Way Trail, I saw a fox squirrel start up a tree after spotting me. It stopped at two feet from the ground when I talked quietly to it. Then it came back down the tree and went leaping off into the underbrush.

I wandered into Far Field Thicket and heard a rustling. There I found a fox sparrow scratching away. I was beginning to see a real pattern here. But I couldn't decide what, if anything, it might mean. To my premodern ancestors, three "foxes" in a day would be a potent omen. But I am content to leave it a mystery—like most seeming signs in nature.

MARCH 12. A light snowfall during the night made for excellent tracking conditions. It looked as if every deer on the mountain had trekked to the churned up, hay-scented fern openings along the powerline right-of-way. As I sat on Turtle Bench, the sun warmed me on one side and the wind cooled me on the other. American crows cawed loudly and continually over First Field and I wondered if they were already moving into a territorial/courting mode.

Sitting on Coyote Bench, I looked down at the peaceful, white-washed, motionless fields of Sinking Valley, still sunk in winter's sleep. I followed fresh woodchuck tracks from the Far Field Road across the field to where their maker had dug out one hole and entered, then exited and proceeded to the burrow system below the barberry shrub, went in and then left and continued on to the old red fox den where it entered and did not exit. No doubt the tracks belonged to a male in search of receptive females.

Older coyote tracks also crossed the field. I cut down to the trail to follow them and then detoured to check out the burrow complex at the edge of Far Field Thicket. There were no tracks there and no sign that any animal was using it.

Coyote tracks wended their way into the woods toward Second Thicket. Others crested the hill and went down the steep mountainside. Still others proceeded into the Far Field woods. Some were very fresh.

Several sets of tracks followed or paralleled Sapsucker Ridge Trail. At one small hole, where a squirrel had dug, fox tracks also approached. Did a fox catch a squirrel? I saw no sign of a battle. Then a set of rabbit tracks crossed the trail. The tracks were as large as a snowshoe hare's. Was it the mythical "mountain rabbit" that one of our neighbors had come to hunt many years ago? He was convinced that the rabbits up here were larger than those in the valley—"mountain rabbits," he called them.

The sunken, U-shaped trenches made by porcupines crossed the trail and proceeded downhill toward Roseberry Hollow. Others headed for the spruce grove and, as usual, I found fresh scat as well as tracks throughout the grove. First, though, at least one porcupine had stopped to climb up a striped maple and nibble the bark all the way to its spindly, top branches.

Back in our yard, I sat down to watch the old apple tree behind the guesthouse that had attracted cedar waxwings, American robins, and even a male downy woodpecker to the bright orange pulp of old, tasteless apples as if they were manna from heaven. Singing dark-eyed juncos, house finches, and black-capped chickadees added color to the scene, but it was the waxwings who clung from branches and pulled out the apples' flesh as the branches swung in the breeze. For a short time, a robin and a waxwing even shared the same apple. The downy perched on the apple itself to dig out the pulp. One waxwing sat frozen in place, yawning once, its yellow belly and reddish-brown head and breast gleaming in the sunlight—birds full of color and life on a cold, white, still winter day. Sometimes they hung upside down to peck at the fruit and extract large beakfuls. Two waxwings shared an apple, but most of the ten waxwings ate individual apples. Even a female house finch perched in one branch, leaned forward, and ate bits of apple. Then she sat on top of the apple itself to eat. Finally a male joined her. Altogether that old tree with its (to us) tasteless apples provided much-needed food to the birds and an hour of easy wildlife watching for me.

Shortly after dinner, as we sat reading in the living room, I heard the cry of more migrating tundra swans. We rushed outside to listen to the whooing cries in the dark, starlit evening.

An hour later, a male great horned owl boomed his bass calls somewhere in the yard and I gazed through the bow window at the moon shining through the trees of Laurel Ridge. Owls and swans and the full moon called me outside again. I pulled on boots and jacket and moved across the yard and into First Field, just as the moon crested the ridge and cast my shadow almost as strongly as the sun. In the moonlight, the snow sparkled, again like it does in strong sunshine. The black-and-white world was almost as luminescent as daylight technicolor.

No more swans called, but the male great horned owl continued his basso profundo love song on Sapsucker Ridge, which echoed off Laurel Ridge. Occasionally, he was answered by the female's alto repetition of his song, but the male was the primary singer during the hour I walked the moonlit First Field, entranced, enthralled, and listening. Then I saw, amid the bright constellations, a single shooting star over Sapsucker Ridge. It provided a final benediction to this white night.

MARCH 13. This date marks the blizzard of 1993, a time I will never forget. At dawn it was snowing hard and throughout the day the thermometer and barometer kept dropping, the wind picked up, and drifts lay against the shed, the veranda, and the guesthouse. For a while we tried to keep the porches open but finally gave up in midafternoon as the birds fought and scratched up whatever they could from the seed I had been continually spreading on the back porch.

Mostly I was on the phone, first with my sister, Linda, then my Dad, and finally Steve, and they all had the same message: This was the blizzard of the century—seventy-mile-an-hour winds; several feet of snow predicted; already five inches in Atlanta, Georgia; cars were being blown off Interstate 95; and Washington, D. C., and Maryland were blacked out. We filled the tub with toilet-flushing water, filled every empty jug with drinking water, gathered together flashlights and kerosene lanterns, and watched the thermometer and barometer continue plummeting. After dinner Dave reported that the barometer was as low as he had ever seen it—28.9.

All night long the wind shook the house, and I remained piled under blankets, cocooned in bed, hoping the electricity would not go or the oil furnaces break down. Surely, I thought, a storm that fierce, which had ground the entire eastern United States to a halt, should convince humans that they don't have the upper hand and never will, that our technological advances are helpless against Nature's most powerful onslaughts, and that at such times, we are all reduced to hope for survival and nothing more.

The next dawn it was ten degrees and the wind still howled, but the snow seemed to have stopped and the birds, some at least, had survived. Bruce cleared the back porch and sprinkled birdseed even before he made our Sunday breakfast of wild blueberry pancakes. Instantly, dozens of birds landed to feed—dark-eyed juncos, seventeen American tree sparrows, four northern cardinals, a couple of house finches, an American goldfinch, song sparrows, black-capped chickadees, white-breasted nuthatches, and tufted titmice. The gang (or most of it) was still here. Somehow those bits of gossamer with beating hearts had survived the long, bitter night. And we still had our electricity and heat, for which we gave thanks.

Frost patterns were etched on every window, which made the house more cavelike. A dim sun shone in a white sky at 8:20 A.M. Whatever it is that pushes the song sparrow's button cannot be warmth. He sang from the juniper tree in the teeth of the frigid wind and cold. One gray squirrel kept returning for birdseed. It was so cold that it moved like an arthritic old man. A female purple finch arrived and sat in the middle of the feeder, along with two goldfinches, to eat seed. Occasional gusts of wind sent clouds of snow swirling up into the air and scattered the birds like chaff, but they were back seconds later. I waded through knee-high drifts as I tried to defend the birds from the more aggressive squirrels. Off in the woods, a deer materialized like some dark phantom.

In early afternoon, Bruce broke trail ahead of me more than a quarter of a mile down our hollow road, and he took photos of the drifts across the road and curving up from the stream. But Dave continued on another mile to the bottom of the road and said the drifts below the big pull-off were like huge sand dunes. He also reported that the main east-to-west rail line at the bottom of our mountain was completely snowed under, and the highway, while cleared, was icy. Only a few SUVs moved slowly along.

By the next day it was two degrees at dawn, clear, still, and warming rapidly to seventeen degrees by 10:00 A.M. Already the eaves dripped in the sunlight. Apparently the jet stream had taken a twist down to the Gulf of Mexico, something that meteorologists can explain but not change, so we had mid-January instead of mid-March temperatures.

Bruce and I snowshoed our way down the hollow road, following Dave's tracks from the day before, but still Bruce had to do more rebreaking and refining and then I added more widening, creating a navigable foot-and-a-half-by-one-foot deep snowshoe track the mile and a half down to our gate. In the hollow only the usual winter songbirds were abroad. Chickadees were particularly visible, working over the still abundant hemlock cones. We also watched a deer wallow down to the stream and up Laurel Ridge. Only deer, squirrel, and mice tracks marred the snow cover.

The drifts at the bottom of the road rolled up the hollow like immobile ocean waves, frozen at the point of breaking, four to eight feet high, and we moved carefully along above the steep slope, sometimes on top of, other times behind a drift. The railroad track had been cleared and the trains were running, but the township road up to the tracks had not been plowed. We snowshoed out to the bridge over the Little Juniata River and watched SUVs and various pieces of snow removal equipment lumber slowly along what still looked like an icy highway.

The hollow was beautiful, especially the still-flowing stream with four or five feet of sculpted, drifted snow along its banks. On our way back up, we stopped at the big pull-off to bask in the sun before snowshoeing, with many rest pauses, back to the house.

This, I reflected then, was probably a once-in-a-lifetime experience, a story we will be telling our grandchildren many years hence. And telling here now because we still mark the date as one of the most memorable natural occurrences we have experienced on our almost thirty-two years on the mountain.

MARCH 14. Thick fog and raining, but both an American robin and a Carolina wren sang in the dawn light. There is a wren in Mexico called the happy wren. Can any wren sound happier than a Carolina wren? Or is any birdsong in late winter more beautiful than a robin's?

A flock of red-winged blackbirds called on Sapsucker Ridge. At the feeders, birds congregated. A woodchuck, looking like a dapper man-about-town, trotted across the snow and down the back slope to his den after what probably had been a big night.

We had a miraculous clearing by early afternoon. Suddenly the fog wafted away, the sun shone, and the March wind whisked off the last of the stormy weather. It was almost fifty degrees at three o'clock and I went out for a short walk. A pair of red-tails flew in tandem, the smaller male above the larger female, and they used the sky over First Field and our yard as a playing field. My heart and spirits soared with theirs, wishing I could be up there with them, enjoying the wind. Then a second male suddenly appeared from nowhere, it seemed, and dove at the first male while the female flew off over Sapsucker Ridge. The two males continued circling and diving and finally disappeared over Laurel Ridge. What was that all about?

I tucked myself into the buttresses of the enormous black oak tree near Turtle Bench and basked in the warm sun. American crows cawed nearby. The ditches ran with snow melt that sparkled in the sunlight. Four turkey vultures sailed past in the wind.

MARCH 15. This morning I stepped outside to hear the low-pitched "fee-bees" of eastern phoebes. On Greenbrier Trail the male brown-headed cowbirds sang their burbling, creaky songs, lifting their wings after each one, while the females inched away from them toward the ends of the tree branches. Lying in First Field, I listened to a singing brown creeper in the woods above and a field sparrow in the field below.

Dave reported hearing the rapid "kik-kik-kik" of a pair of Cooper's hawks up in the woods most of the day. He also reported that the first garter snake was out near the guesthouse stone wall. And the male kestrel was back on its favorite electric line near the barn.

By afternoon the temperature was in the mid-seventies. Spring had arrived on the Ides of March.

MARCH 16. A lovely sunrise this partially overcast, warm day. I awoke to a chorus of song sparrows, each with a slightly different rendition of their standard "Hip! Hip! Hurrah! boys, Spring is here" song. And the field sparrow sang its melodious, descending trill.

The ducklike quacks of wood frogs drew me down to the pond. After I'd sat still at the edge of the tiny pond for over

half an hour, nine males eventually surfaced and started swim-
ming and calling. A large, pink female, bulging with eggs,
emerged from the pond bottom and was instantly grabbed by
a male. Two others tried, half-heartedly, to interfere, but the
pair kicked them off and dove down into the water out of
sight. By then the sky had clouded over and it started to
sprinkle, plus my one foot was numb. Finally, I couldn't put
off flexing it any longer. The instant I moved my foot, all the
frogs dove under the water. It took several minutes for feeling
to return to the foot so I could walk away. Already a small
clump of eggs was attached to a large blade of grass lying in
the pond.

At the pond I had entertained myself during slow minutes
by watching water striders, which overwinter as adults under
rocks or logs at the bottom of streams or ponds. They too are
engaged in courtship and mating practices. A male water
strider locates a good position near a piece of vegetation or
log in the water and stays there making ripple patterns by
moving his middle legs up and down rapidly, probably to
keep off other males and to attract a receptive female. She
touches the male with either her leg or mouth and then allows him to mate while she holds on to an egg-laying object.
Then she lays her eggs there while the male remains nearby
guarding her from other males. I saw the first part of that action, only I had assumed that the come-hither creature was
the female. I also saw one water strider on top of another but
no egg-laying.

First Field was alive with song, but the tufted titmice engaged in an aggressive ballet of their own. Three followed
each other up and down tree branches while calling. Once
two of them tumbled earthward in a swirl, feet locked together, while a third tried to break them apart. Then two of
them sang back and forth as the third one looked on silently.
Apparently, the two males had been engaged in a territorial
battle that had turned into a singing fest for the benefit of the
watching female.

On this date, in 1996, our son Mark had called from Tegucigalpa, Honduras, to tell us that our first grandchild, a girl, had been born. Today Eva Luz celebrated her fifth birthday and we attended a family party for her in midafternoon.

She does so love a party, even a sedate one like ours. But she danced for us, as she often does, and was widely pleased with her birthday cake and presents. And with Poppop, who always enjoys watching her antics.

MARCH 17. Twenty-eight degrees and snowing at dawn. We were back to winter, which stunned most of the birds into silence except for the Carolina wrens that caroled in the dawn light. Hundreds of blackbirds, mostly common grackles, but a few European starlings, brown-headed cowbirds, and red-winged blackbirds, fell like black rain on the back porch, feeders, steps, and slope, blanketing the area while the feeder regulars fled. The blackbirds swept in and out for twenty minutes or more before most took off. But a couple of cowbirds and red-winged blackbirds remained to mingle with a fox sparrow, American goldfinch, and the usual feeder regulars.

A few snowflakes continued to flutter down in the almost complete silence along Greenbrier Trail. Then, as I sat along the stream, the tick of sleet began. But it stopped again by the time I reached the forks. A veritable chorus of birds sang on this gloomy day, mostly our resident songbirds, but they were joined by a field sparrow.

Large snowflakes fell in early afternoon—a brief "onion snow." Then again the tick of sleet against the windows as fog sifted through the trees and enshrouded the ridge tops. A mix of freezing rain, fist-sized snowflakes, and rain continued to fall off and on, finally settling into freezing rain.

MARCH 18. Mostly clear and windy as March, the seesaw month, slowly warmed again. Song sparrows and Carolina wrens made up the chilly dawn chorus, but at 9:15 A.M., as I headed across First Field to Big Tree Trail, I heard a singing eastern bluebird and eastern phoebe from the warm, open corner of the field. While sitting at the base of a big cherry tree at the edge of First Field, the cold wind thundered around me, but I was warmed by the sun on this exhilarating March day.

As I walked across the field, a shadow passed over the snow, and I looked up to see a silent flock of ring-billed gulls floating low overhead. Their white bodies with black wing tips shone like apparitions against the blue sky. They were so close I could even see the black ring on their bills. Already they were heading northwest for the Great Lakes, where they breed and raise their young.

It was warm enough that the wood frogs were calling again when I returned from my walk. Now there were six clumps of eggs and fewer callers than the other day. But I went back in midafternoon and crawled through the dried weeds of the field, easing myself up over the hillock in front of the tiny, manmade pond. After a few minutes the wood frogs spotted me and dove out of sight, so once again I sat down in front of the pond without moving for a half hour until one by one froggy heads appeared. Finally they began calling and bumping into each other. All thirty-five were males during the forty-five minutes I watched.

Who knows what magic they evoke to pull me down to the pond to watch in the breezy brilliant sunlight? But I love their froggy shapes, particularly their long legs that propel them so elegantly through the water. Then, too, the setting is evocative—the still-dried beige stalks of weeds and grasses matted around the pond, the "coo" of a mourning dove, the soughing of wind in the white pine trees nearby, the sparkling stream water, the singing of song sparrows, black-capped chickadees, and eastern phoebes. All speak of the first stirrings of spring sweetened by the sound and sight of male wood frogs calling in the females for their annual mating ritual. And, as I walked back up our road, I found the bright yellow disks of coltsfoot growing beside the driveway—the first flower of spring.

MARCH 19. A hazy-clear dawn. I was awakened by the liquid-burble of brown-headed cowbird song. Later, as I walked the Laurel Ridge Trail, the valley was still wrapped in fog even though we had been suffused in sunlight since sunrise. Canada geese streamed overhead, their cries drowning out the limestone quarry bedlam in the valley.

At the Far Field black-capped chickadees duetted. One sang "pee-wee" and another answered, picking up the pitch of the first one's "wee" with its "pee" and then scaling its "wee" down the same number of notes as the first bird's "pee-wee." Altogether there were two birds with higher-pitched "pee-wees" and one with a lower-pitched song.

Buffeted by the wind at the top of First Field, I stood beneath a panorama of blue sky studded with wispy clouds and both heard and saw a wavering "V" of one hundred geese flying high overhead. A Cooper's hawk followed by a turkey vulture sailed past.

The air smelled of melting spring warmth even as a sudden March wind picked up, drowning out the intrusive valley sounds. As I lay in the field, clouds loomed up from the south, quickly engulfing the sun, and foretelling the threatened afternoon storm, the geese having flown just ahead of the front.

I surprised a woodchuck in the middle of the field and it ran into its burrow. As I walked toward it, the woodchuck popped its head out to look around and I froze in place. Up and down like a periscope its head went as it slowly looked in every direction for many long minutes while I got colder and colder. The wind blew, clouds blanketed the sky, and a pair of red-tailed hawks soared above the ridge as I sat watching the woodchuck from twenty-five feet away.

Finally, after many "up periscope" looks, it eased most of its body out of the burrow but did not resume feeding. By then I was chilled to the bone because the temperature had dropped many degrees. I walked deliberately toward the woodchuck, and it dove down into the hole. As I reached the burrow entrance, loud noises, like a horse whinnying, came from it.

Heavy rains and wind continued all afternoon and evening and we were back into winter again.

MARCH 20. Spring officially arrived at 2:58 this afternoon. But it was a disappointing day because it was raining hard and fog blotted out everything beyond our driveway. At 6:40 P.M., when Bruce headed down to our compost heap with the day's garbage, he heard the "peenting" call of an American woodcock near our barn. He ran back to tell Dave and me, and we rushed down to listen and watch in the light mist and dusky light.

The show continued for twenty minutes. Sometimes the woodcock flew directly overhead, but we couldn't see anything except an occasional flash of wings, because it was almost dark and the clouds were below the mountaintop. The number of "peents" varied from five to seven to twelve to twenty-seven before we heard him chirping above us during his song flight, and then finally the whistling of his wings as he plummeted to earth. The "peents" seemed to come from several directions, even overhead, yet woodcock males are supposed to emit them only on the ground. Researchers claim that the displaying male rotates on the ground, which causes a directional change in the intensity of "peents." That may explain what we heard or, possibly, that this woodcock had his own ideas about how to proceed.

Many eastern North American fields are used as singing grounds during the spring, but the functions of the peenting and chirping song flights have not been studied. Ornithologists assume that they are used to advertise the position of each bird to other woodcocks, but they often peent and fly when they are alone.

However, if another male appears or a female visits, the display is intensified. Up to six males have been counted at one singing ground and, as they migrate north, they move from singing ground to singing ground where they mate with whatever females they find. Most continue displaying and mating for two months, once they arrive on their breeding grounds, which can be as far north as southern Canada.

They display twice a day, at dawn and dusk. At dusk they fly or walk to their singing ground from wherever they have spent their day and then, after their display, they fly to a separate place to spend the night or sometimes remain on the singing ground overnight. At dawn they fly back to their singing ground. Probably the length of their displays is triggered by light intensity because, on foggy evenings, like this one, they begin and end earlier.

Over the years our First Field has been only an occasional stopover point for a migrating American woodcock. Most years we neither see nor hear any. To have both heard and seen one on this last day of winter and first day of spring bestowed a blessing on the season past and kindled expectation for the season ahead.

And so I end this last of my four seasons books at the beginning of a new century. I agree with William Browne, who wrote four centuries ago: "There is no season such delight can bring as summer, autumn, winter and the spring," especially on a central Appalachian mountaintop. As I enter my seventh decade, I want to spend even more time outside recording the natural world.

Changes have come to our mountain over the three decades we have lived here and most have reflected a world overwhelmingly dominated by the human race. An interstate highway has scarred many miles of our once-intact mountain ridge and will gobble up more interior forest land. New shopping malls are to be gouged into the side of this same ridge to service the people traveling on the interstate. What was once forty miles of relatively unsettled mountain land will be fragmented by these and other planned inroads.

Similar scenarios are occurring throughout our state and indeed throughout our nation and the rest of the world. As the biological diversity of the natural world continues to disappear, I hope that my books and the notes I have taken over the years will leave a record for those to come of what we once had here and what we are now selling for a pittance. Our grandchildren and great-grandchildren will not thank us for our thoughtlessness and greed as we pave over more and more of what was once a splendid gift of grace and beauty.

Living here has been both a blessing and a curse: A blessing because of what I have seen and learned and felt; a curse because I can remember what it used to be. Now our air is more heavily polluted, our peace has been shattered, the wildlife is diminishing, and our forest is not regenerating as it should.

And yet . . . spring has come again. The long, white winter is over. If watching the seasons closely has taught me anything, it is that the natural world will continue, in some form or another, long after humanity has left the scene. What a pity, though, that there will be no human eyes to admire this splendid world we have been given.

Or just maybe, in this new century, we will change our rapacious ways and learn to appreciate what is here before it is too late. I hope so. I desperately want my granddaughter and her children and their children to live in a world of intact natural systems, able to have rich, satisfying lives that include time in the outdoors.

The white magic of another winter has ended and once again the promise of spring is in the air. To all of you who have walked with me through an Appalachian year, thank you and farewell. May you enjoy many more springs, summers, autumns, and winters on this wondrous part of the earth that we call the Appalachian Mountains.

Selected Bibliography

General

Bass, Rick. *Winter: Notes from Montana*. Boston: Houghton Mifflin, 1991.

Bellaby, Mara D. "Extinction Risk Grows across the Globe." *Associated Press Newswires* (September 28, 2000).

Bentley, W. A., and W. J. Humphreys. *Snow Crystals*. New York: Dover Publications, 1962.

Blanchard, Duncan C. *The Snowflake Man: A Biography*. Granville, Ohio: McDonald & Woodward Publishing Co., 1998.

_____. "The Snowflake Man." *Weatherwise* 23, no. 6: 260–69.

Bolgiano, Chris. *The Appalachian Forest: A Search for Roots and Renewal*. Mechanicsburg, Pa.: Stackpole Books, 1998.

_____. *Living in the Appalachian Forest: True Tales of Sustainable Forestry*. Mechanicsburg, Pa.: Stackpole Books, 2002.

Bonta, Mark. *Bioplum: A Biological Inventory of the Bonta Property and Adjacent Parcels Northern Brush Mountain, Blair County, Pennsylvania*. Plummer's Hollow, Pa.: 1991.

Brown, Lauren. *Weeds in Winter*. Boston: Houghton Mifflin, 1976.

Brooks, Maurice. *The Appalachians*. Boston: Houghton Mifflin, 1965.

Fergus, Charles. *Wildlife in Pennsylvania and the Northeast*. Mechanicsburg, Pa.: Stackpole Books, 2000.

Griffin, Steven A. *Snowshoeing*. Mechanicsburg, Pa.: Stackpole Books, 1998.

Heinrich, Bernd. *Winter World: The Ingenuity of Animal Survival*. New York: HarperCollins Publishers, Inc., 2003.

Kingsolver, Barbara. *The Poisonwood Bible*. New York: Harper-Flamingo, 1998.

Knight, Charles, and Nancy Knight. "Snow Crystals." *Scientific American* 228 (January 1973): 103–7.

Lyon, Thomas J. *This Incomparable Land: A Guide to Nature Writing*. Minneapolis: Milkweed Editions, 2001.

Marchand, Peter J. *Life in the Cold: An Introduction to Winter Ecology.* Hanover, N.H.: University Press of New England, 1991.

Marigo, Luiz Claudio. "Luiz Claudio Marigo." *Natural History* 109 (December 2000): 100.

Murie, Olaus J. *A Field Guide to Animal Tracks.* Boston: Houghton Mifflin, 1954.

Nabhan, Gary Paul, and Stephen Trimble. *The Geography of Childhood: Why Children Need Wild Places.* Boston: Beacon Press, 1994.

Norris, Kathleen. *The Cloister Walk.* New York: G. P. Putnam's Sons, 1996.

Osgood, William E., and Leslie Hurley. *The Snowshoe Book.* Brattleboro, Vt.: Greene Press, 1975.

Sanders, Scott R. *Hunting for Hope: A Father's Journey.* Boston: Beacon Press, 1998.

Smith, Richard P. *Animal Tracks and Signs of North America.* Mechanicsburg, Pa.: Stackpole Books, 1982.

Stokes, Donald W. *A Guide to Nature in Winter.* Boston: Little, Brown, 1976.

Taubes, Gary. "The Snowflake Enigma." *Discover* 5 (January 1984): 75–78.

Teale, Edwin Way. *Wandering through Winter.* New York: Dodd, Mead & Co., 1965.

Thoreau, Henry D. *Faith in a Seed.* Washington, D.C.: Island Press, 1993.

Todd, Kim. *Tinkering with Eden: A Natural History of Exotics in America.* New York: W. W. Norton & Co., 2001.

Torrey, Bradford, and Francis H. Allen, ed. *The Journal of Henry D. Thoreau.* 2 vols. New York: Dover Publications, 1962.

Weidensaul, Scott. *Mountains of the Heart: A Natural History of the Appalachians.* Golden, Colo.: Fulcrum Publishing, 1994.

Birds

Bent, Arthur Cleveland. *Life Histories of North American Birds of Prey.* 2 vols. New York: Dover Publications, 1961.

_____. *Life Histories of North American Cardinals, Grosbeaks, Buntings, Towhees, Finches, Sparrows and Their Allies.* 3 vols. New York: Dover Publications, 1968.

_____. *Life Histories of North American Gallinaceous Birds.* Washington, D.C.: Smithsonian Institution, 1932.

_____. *Life Histories of North American Nuthatches, Wrens,*

Thrashers, and Their Allies. New York: Dover Publications, 1964.

_____. *Life Histories of North American Jays, Crows, and Titmice.* New York: Dover Publications, 1964.

_____. *Life Histories of North American Thrushes, Kinglets, and Their Allies.* New York: Dover Publications, 1964.

_____. *Life Histories of North American Woodpeckers.* New York: Dover Publications, 1964.

Bildstein, Keith L., and Ken Meyer. "Sharp-Shinned Hawk." *The Birds of North America* 482 (2000): 1–27.

Brauning, Daniel W., ed. *Atlas of Breeding Birds in Pennsylvania.* Pittsburgh: University of Pittsburgh Press, 1992.

Brawn, Jeffrey D., and Fred B. Samson. "Winter Behavior of Tufted Titmice." *Wilson Bulletin* 95 (June 1983): 222–32.

Bull, Evelyn L., and Jerome A. Jackson. "Pileated Woodpecker." *The Birds of North America* 148 (1995): 1–20.

Galati, Robert. *Golden-Crowned Kinglets: Treetop Nesters of the North Woods.* Ames: Iowa State University Press, 1991.

Ghalambor, Cameron K., and Thomas E. Martin. "Red-Breasted Nuthatch." *The Birds of North America* 459 (1999): 1–27.

Grubb, Thomas C., Jr., *Tufted Titmouse.* Mechanicsburg, Pa.: Stackpole Books, 1998.

Grubb, Thomas C., Jr., and V. V. Pravosudov. "Tufted Titmouse." *The Birds of North America* 86 (1994): 1–15.

Halkin, Sylvia L., and Susan U. Linville. "Northern Cardinal." *The Birds of North America* 440 (1999): 1–31.

Haggerty, Thomas M., and Eugene S. Morton. "Carolina Wren." *The Birds of North America* 188 (1995): 1–18.

Hamerstrom, Frances. *Harrier, Hawk of the Marshes: The Hawk That Is Ruled by a Mouse.* Washington, D.C.: Smithsonian Institution Press, 1986.

Heinrich, Bernd. *One Man's Owl.* Princeton, N.J.: Princeton University Press, 1987.

Hendricks, Paul. "Ground-Caching and Covering of a Food by a Red-Breasted Nuthatch." *Journal of Field Ornithology* 66 (Summer 1995): 370–72.

Houston, C. Stuart, Dwight G. Smith, and Christoph Rohner. "Great-Horned Owl." *The Birds of North America* 372 (1998): 1–27.

Ingold, J. L., and Robert Galati. "Golden-Crowned Kinglet." *The Birds of North America* 301 (1997): 1–24.

Jackson, Jerome A., and Henri R. Ouellet. "Downy Woodpecker." *The Birds of North America* 613 (2002): 1–31.

Jackson, Jerome A., Henri R. Ouellet, and Bette J. S. Jackson. "Hairy Woodpecker." *The Birds of North America* 702 (2002): 1–27.

Kilham, Lawrence. *Woodpeckers of Eastern North America.* New York: Dover Publications, 1992.

Limpert, R. J., and S. L. Earnst. "Tundra Swan." *The Birds of North America* 89 (1994): 1–19.

Macwhirter, R. Bruce, and Keith L. Bildstein. "Northern Harrier." *The Birds of North America* 210 (1996): 1–31.

McWilliams, Gerald M., and Daniel W. Brauning. *The Birds of Pennsylvania.* Ithaca, N.Y.: Cornell University Press, 2000.

Marks, J. S., D. L. Evans, and D. W. Holt. "Long-Eared Owl." *The Birds of North America* 133 (1994): 1–23.

Middleton, Alex L. *American Goldfinch.* Mechanicsburg, Pa.: Stackpole Books, 1998.

Naugler, Christopher T. "American Tree Sparrow." *The Birds of North America* 37 (1993): 1–11.

Peterson, Roger Tory. *A Field Guide to the Birds of Eastern and Central North America.* Boston: Houghton Mifflin, 2002.

Pravosudov, V. V., and T. C. Grubb, Jr. "White-Breasted Nuthatch." *The Birds of North America* 54 (1993): 1–14.

Preston, C. R., and R. D. Beane. "Red-Tailed Hawk." *The Birds of North America* 52 (1993): 1–23.

Preston, Charles R. *Red-Tailed Hawk.* Mechanicsburg, Pa.: Stackpole Books, 2000.

Ritchison, Gary. *Downy Woodpecker.* Mechanicsburg, Pa.: Stackpole Books, 1999.

_____. *Northern Cardinal.* Mechanicsburg, Pa.: Stackpole Books, 1997.

Shackelford, Clifford E., Raymond E. Brown, and Richard N. Conner. "Red-Bellied Woodpecker." *The Birds of North America* 500 (2000): 1–23.

Smith, Dwight G. *Great Horned Owl.* Mechanicsburg, Pa.: Stackpole Books, 2002.

Smith, Susan. "Black-Capped Chickadee." *The Birds of North America* 39 (1993): 1–18.

_____. *The Black-Capped Chickadee: Behavioral Ecology and Natural History.* Ithaca, N.Y.: Comstock Publishing Association, 1991.

_____. *Black-Capped Chickadee.* Mechanicsburg, Pa.: Stackpole Books, 1997.

Stokes, Donald W. *A Guide to Bird Behavior: Volume I.* Boston: Little, Brown, 1979.

Stokes, Donald W., and Lillian Q. Stokes. *A Guide to Bird Behavior: Volume II.* Boston: Little, Brown, 1983.

———. *A Guide to Bird Behavior: Volume III.* Boston: Little, Brown, 1989.

Wink, J., S. E. Senner, and L. J. Goodrich. "Food Habits of Great Horned Owls in Pennsylvania." *Proceedings of the Pennsylvania Academy of Sciences* 61, no. 2: 133–37.

Winn, Marie. *Red-Tails in Love.* New York: Random House Inc., 1999.

Witmer, M. C., D. J. Mountjoy, and L. Elliot. "Cedar Waxwing." *The Birds of North America* 309 (1997): 1–27.

Wood, Merrill. *Birds of Central Pennsylvania.* State College, Pa.: State College Bird Club, 1976.

Yahner, Richard H., and Thomas E. Morrell. "Hunters of the Night." *Pennsylvania Game News* 64 (February 1993): 33–35.

Insects

Berenbaum, May R. *Bugs in the System: Insects and Their Impact on Human Affairs.* Reading, Mass.: Addison-Wesley, 1995.

Borror, Donald J., and Richard E. White. *A Field Guide to the Insects of America North of Mexico.* Boston: Houghton Mifflin, 1970.

Covell, Charles V., Jr., *Eastern Moths.* Boston: Houghton Mifflin, 1984.

Heinrich, Bernd. "Some Like It Cold." *Natural History* 103 (February 1994): 42–49.

Mitchell, Robert T., and Herbert S. Zim. *Butterflies and Moths: A Guide to the More Common American Species.* New York: Golden Press, 1964.

Opler, Paul A., and Vichai Malikul. *Eastern Butterflies.* Boston: Houghton Mifflin, 1992.

Stokes, Donald W. *A Guide to Observing Insect Lives.* Boston: Little, Brown, 1983.

Mammals

Althoff, Donald P., Gerald L. Storm, and David R. DeWalle. "Daytime Habitat Selection by Cottontails in Central Pennsylvania." *Journal of Wildlife Management* 61 (1997): 450–59.

Boer, Arnold H. *Ecology and Management of the Eastern Coyote.* Fredericton, New Brunswick: Wildlife Research Unit, University of New Brunswick, 1992.

Chapman, Joseph A., J. Gregory Hockman, and Magaly M. Ojeda C. "Sylvilagus floridanus." *Mammalian Species* 136 (April 15, 1980): 1–8.

Churchill, Sara. *The Natural History of Shrews.* London: Christopher Helm, 1990.

Dolan, Patricia G., and Dilford C. Carter. "Glaucomys volans." *Mammalian Species* 78 (June 15, 1977): 1–6.

Fritzell, Erik K., and Kurt J. Haroldson. "Urocyon cinereoargenteus." *Mammalian Species* 189 (November 23, 1982): 1–8.

Gehrt, Stanley D., and Erik K. Fritzell. "Behavioral Aspects of the Raccoon Mating System: Determinants of Consortship Success." *Animal Behavior* 57 (March 1999): 593–601.

Gilbert, Bil. "A Groundhog's Day Means More to Us Than It Does to Him." *Smithsonian* 15 (February 1985): 60–69.

———. "Coyotes Adapted to Us, Now We Have to Adapt to Them." *Smithsonian* 21 (March 1991): 69–79.

Gilmore, Robert M., and J. Edward Gates. "Habitat Use by the Southern Flying Squirrels at a Hemlock-Northern Hardwood Ecotone." *Journal of Wildlife Management* 49, no. 3: 703–10.

Gurnell, John. *The Natural History of Squirrels.* New York: Facts on File, 1987.

Hamilton, William J., Jr., and John O. Whitaker, Jr. *Mammals of the Eastern United States.* Ithaca, N.Y.: Cornell University Press, 1998.

Hayden, Arnold H. "The Eastern Coyote Revisited." *Pennsylvania Game News* 60 (December 1989): 12–15.

Holmgren, Virginia C. *Raccoons in History, Folklore, and Today's Backyards.* Santa Barbara, Calif.: Capra Press, 1990.

Kinkead, Eugene. *Squirrel Book.* New York: E.P. Dutton, 1980.

Lotze, Joerg-Henner, and Sydney Anderson. "Procyon lotor." *Mammalian Species* 119 (June 8, 1979): 1–8.

Merritt, Joseph F. "Clethrionomys gapperi." *Mammalian Species* 146 (May 8, 1981): 1–9.

———. *Guide to the Mammals of Pennsylvania.* Pittsburgh: University of Pittsburgh Press, 1987.

———. "Winter Survival Adaptations of the Short-Tailed Shrew *(Blarina Brevicauda)* in an Appalachian Montane Forest." *Journal of Mammalogy* 67 (1986): 450–64.

Merritt, Joseph F., and David A. Zegers. "Seasonal Thermogenesis and Body-Mass Dynamics of *Clethrionomys gapperi.*" *Canadian Journal of Zoology* 69 (1991): 2771–77.

Parker, Gerry. *Eastern Coyote: The Story of Its Success.* Halifax, Nova Scotia: Nimbus Publishing Limited, 1995.

Roze, Uldis. *The North American Porcupine.* Washington, D.C.: Smithsonian Institution Press, 1989.

Ryden, Hope. *God's Dog: A Celebration of the North American Coyote.* New York: Lyons & Burford, 1975.

Smith, Charles W. G. "Yellow Eyes." *Country Journal* 19 (March/April 1992): 26–31.

Stains, H. J. "The Raccoon in Kansas: Natural History, Management, and Economic Importance." *Miscel. Publ. Museum Natural History University of Kansas* 10 (1956): 1–76.

Steele, Michael A., and John L. Kroprowski. *North American Tree Squirrels.* Washington, D.C.: Smithsonian Institution Press, 2001.

Wells-Gosling, Nancy. *Flying Squirrels: Gliders in the Dark.* Washington, D.C.: Smithsonian Institution Press, 1985.

Wishner, Lawrence. *Eastern Chipmunks: Secrets of Their Solitary Lives.* Washington, D.C.: Smithsonian Institution Press, 1982.

Yahner, Richard H. *Fascinating Mammals: Conservation and Ecology in the Mid-Eastern States.* Pittsburgh: University of Pittsburgh Press, 2001.

Zeveloff, Samuel I. *Raccoons: A Natural History.* Washington, D.C.: Smithsonian Institution Press, 2002.

Plants

Fergus, Charles. *Trees of Pennsylvania and the Northeast.* Mechanicsburg, Pa.: Stackpole Books, 2002.

Grimm, William Carey. *The Shrubs of Pennsylvania.* Harrisburg, Pa.: Stackpole and Heck, 1952.

———. *The Trees of Pennsylvania.* Harrisburg, Pa.: Stackpole and Heck, 1950.

Heinrich, Bernd. *The Trees in My Forest.* New York: HarperCollins Publishers, 1997.

Kimmerer, Robin Wall. *Gathering Moss: A Natural and Cultural History of Mosses.* Corvallis: Oregon State University Press, 2003.

Little, Charles E. *The Dying of the Trees: The Pandemic in America.* New York: Viking Penguin, 1995.

Peattie, Donald Culross. *A Natural History of Trees of Eastern and Central North America.* 2nd edition. New York: Bonanza Books, 1966.

Purvis, William. *Lichens.* Washington, D.C.: Smithsonian University Press, 2000.

Shuttleworth, Floyd S., and Herbert S. Zim. *Non-Flowering Plants.* New York: Golden Press, 1967.

Yahner, Richard H. *Eastern Deciduous Forest: Ecology and Wildlife Conservation*. Minneapolis: University of Minnesota Press, 1995.

Reptiles and Amphibians

Conant, Roger, and Joseph T. Collins. *A Field Guide to Reptiles and Amphibians East and Central North America*. Boston: Houghton Mifflin, 1991.

Green, N. Bayard, and Thomas K. Pauley. *Amphibians and Reptiles in West Virginia*. Pittsburgh: University of Pittsburgh Press, 1987.

Index